A Comparative Study of Chinese and Western Cyclic Myths

Asian Thought and Culture

Charles Wei-hsun Fu
General Editor

Vol. VIII

PETER LANG
New York • San Francisco • Bern • Baltimore
Frankfurt am Main • Berlin • Wien • Paris

Robert Shanmu Chen

A Comparative Study of Chinese and Western Cyclic Myths

PETER LANG
New York • San Francisco • Bern • Baltimore
Frankfurt am Main • Berlin • Wien • Paris

Library of Congress Cataloging-in-Publication Data

Chen, Robert Shanmu
  A comparative study of Chinese and Western cyclic myths / Robert Shanmu Chen.
    p. cm. — (Asian thought and culture ; vol. 8)
    Includes bibliographical references and index.
    1. Time—Mythology. 2. Eternal return. 3. Mythology, Chinese.
  4. Myth in literature. I. Title. II. Series.
  BL325.T55C44   1992   115—dc20   91-21217
  ISBN 0-8204-1675-4   CIP
  ISSN 0893-6870

Die Deutsche Bibliothek-CIP-Einheitsaufnahme

Chen, Robert Shanmu:
A comparative study of Chinese and western cyclic myths / Robert Shanmu Chen.—New York; Berlin; Bern; Frankfurt/M.; Paris; Wien; Lang, 1992
  (Asian thought and culture ; Vol. 8)
  ISBN 0-8204-1675-4
NE: GT

The paper in this book meets the guidelines for permanence and durability of the Committee on Production Guidelines for Book Longevity of the Council on Library Resources.

∞

© Peter Lang Publishing, Inc., New York 1992

All rights reserved.
Reprint or reproduction, even partially, in all forms such as microfilm, xerography, microfiche, microcard, offset strictly prohibited.

Printed in the United States of America.

ACKNOWLEDGMENTS

I am very grateful to the following authors and publishers for permission to quote from copyrighted materials:

Excerpt from *A Brief History of Chinese Fiction* by Lu Hsün, trans. by Hsien-yi Yang and Gladys Yang, Copyright© 1958 by Foreign Languages Press, reprinted by permission of Foreign Languages Press.

Excerpt from *Anger and After: A Guide to the New British Drama* by John Russel Taylor, Copyright© 1963 by Methuen & Co. (London), reprinted by permission of Octopus Publishing Group Library for the world market.

Excerpts from *A Short History of Chinese Philosophy* by Fung Yu-lan, Copyright© 1966 by The Free Press, reprinted by permission of MacMillan Publishing Co.

Excerpt from *Beckett* by Richard Coe, Copyright© 1964 by Oliver and Boyd, reprinted by permission of Longman Group UK Ltd.

Excerpts from "Chinese Concept of Time" by Nathan Sivin in *The Earlham Review* I (Fall, 1966), Copyright© 1966 by *The Earlham Review*, reprinted by permission of *The Earlham Review*.

Excerpts from *Comparative Literature: Method and Perspective* by Newton Stallknecht and Horst Frenz (editor), Copyright© 1971 by Southern Illinois University Press, reprinted by permission of Southern Illinois University Press.

Excerpt from *Ch'u Tz'u: The Songs of the South* by David Hawks, Copyright© 1962 by Beacon Press, reprinted by permission of the author.

Excerpt from *Dante: A Collection of Critical Essays* by John Freccero (editor), Copyright© 1965 by Prentice-Hall Inc., reprinted by permission of Les Presses De La Cité.

Excerpt from *Dante: Poet of the Secular World* by Erich Auerbach, trans. by Ralph Mannhaim, Copyright© 1961 by The University of Chicago Press, reprinted by permission of The University of Chicago Press.

Excerpt and illustration of Dante's Cosmos and Hell from *Dante's Divine Comedy* by Thomas Bergin, Copyright© 1971 by Prentice-Hall Inc., reprinted by permission of Prentice-Hall Inc.

Excerpts from *Fables of Identity* by Northrop Frye, Copyright© 1963 by Harcourt, Brace, and World, reprinted by permission of Harcourt Brace Jovanovich, Inc.

Excerpt from *Faust: Part Two* by Goethe, trans.by Philip Wayne (Penguin Classics, 1959), Copyright© 1959 by the Estate of Philip Wayne, reprinted by permission of Penguin Books Ltd.

Excerpt from *Gertrude Stein and the Literature of the Modern Consciousness* by Norman Weinstein, Copyright© 1970 by Ungar Publishing Co., reprinted by permission of Ungar Publishing Co.

Excerpt from *Intellectual Foundation of China* by Frederick Mote, Copyright© 1971 by Alfred Knopf Publishing, reprinted by permission of Random House, Inc.

Excerpts from *Inventory* by Michel Butor, trans. by Richard Howard, Copyright© 1968 by Simon and Schuster, reprinted by permission of Georges Borchardt.

Excerpts from *Language and Silence* by George Steiner, Copyright© 1967 by Atheneum, reprinted by permission of MacMillan Publishing Co. for U.S., its territories and dependencies and Canada; and by permission of Faber and Faber Limited Publishers for the world market.

Excerpt from *Leonard Cohen* by Michael Ondaatje, Copyright© 1970 by McClelland & Stewart, reprinted by permission of McClelland & Stewart.
Excerpts from "Leonard Cohen: Black Romantic," by Sandra Djwa, in *Canadian Literature*, (34) 1967, reprinted by permission of the author.
Excerpts from *Moll Flanders* by Deniel Defoe, ed. by James Sutherland, Riverside Edition, Copyright© 1959 by Houghton Mifflin Company. Used with permission.
Excerpts from *Myth and Literature: Contemporary Theory and Practice* by John B. Vickery, (editor), Copyright© 1966 by University of Nebraska Press, reprinted by permission of University of Nebraska Press.
Excerpts from *Mythopoesis: Mythic Patterns in the Literary Classics* by Harry Slochower, Copyright© 1970 by Wayne State University Press, reprinted by permission of Wayne State University Press.
Excerpt from *Odysseus Ever Returning: Essays on Canadian Writers and Writing* by George Woodcock, Copyright© 1970 by McClelland & Stewart, reprinted by permission of the author.
Excerpts from *Pamela* by Samuel Richardson, Copyright© 1959 by J. M. Dent & Sons Ltd. reprinted by permission of David Campbell Publishers Ltd. from Samuel Richardson *Pamela* in the Everyman's Library edition.
Excerpts from *Philosophy in a New Key* by Susanne Langer, Copyright© 1951 by Harvard University Press, reprinted by permission of Harvard University Press, Copyright© 1942, 1951, 1957 by the President and Fellows of Harvard College,© renewed 1970, 1979 by Susanne Knauth Langer; 1985 by Leonard C. R. Langer.
Excerpt from "Rhymeprose on Literature: The Wen-fu of Lu Chi," by Achilles Fang, in *Harvard Journal of Asiatic Studies*, 14(1951), Copyright© 1951 by *Harvard Journal of Asiatic Studies*, reprinted by permission of *Harvard Journal of Asiatic Studies*.
Excerpts from *Self and Society in Ming Thought*, ed. by Wm. Theodore de Bary, Copyright© 1970 by Columbia University Press, reprinted by permission of Columbia University Press.
Excerpts from *Studies in Human Time* by Georges Poulet, Copyright© 1956 by Johns Hopkins University Press, reprinted by permission of Johns Hopkins University Press.
Excerpt from *The Absurd: The Critical Idiom* by Arnold Hinchcliff, Copyright© 1967 by Methuen & Co., reprinted by permission of Methuen & Co.
Excerpt from *The Blue and Brown Books* by Ludwig Wittgenstein, Copyright© 1958 by Blackwell Publishers, Ltd., reprinted by permission of Blackwell Publishers.
Excerpts from *The Chinese Mind* by Charles Moore, Copyright© 1967 by University of Hawaii Press, reprinted by permission of University of Hawaii Press.
Excerpts from *The Classic Chinese Novel* by C. T. Hsia, Copyright© 1968 by Columbia University Press, reprinted by permission of the author.
Excerpts from *The Homecoming* by Harold Pinter, Copyright© 1965 by Grove Press, reprinted by permission of Faber and Faber Limited Publishers.
Excepts from *The Hsi yu chi: A Study of Antecedents to the Sixteenth Century Chinese Novel* by Glen Dudbridge, Copyright© 1970 by Cambridge University Press, reprinted by permission of Cambridge University Press.
Excerpts from *The Norton Anthology of Poetry* by Arthur Eastman (editor), Copyright© 1970 by W. W. Norton & Company, Inc., reprinted by permission of W. W. Norton & Company, Inc.

Excerpts from *Theory of Literature* by Rene Wellek and Austin Warren, Copyright© 1949 by Harcourt, Brace and Co., reprinted by permission of Harcourt, Brace Jovanovich, Inc.

Excerpt from *The Secret of the Golden Flower* by Richard Wilhelm, (Arkana, 1984), Copyright© 1984 by Richard Wilhelm, reprinted by permission of Penguin Books Ltd.

Excerpts from *The Soul of China* by Amaury de Riencourt, Copyright © 1958 by Coward-McCaun Inc., reprinted by permission of the author.

Excerpts from *The Theatre of Revolt* by Robert Brustein, Copyright© 1964 by Atlantic Monthly Press, reprinted by permission of American Repertory Theatre.

Excerpts from *Time and Eastern Man* by Joseph Needham, Royal Anthropological Institute Occasional Paper 21 (1965), Copyright© 1965 by Royal Anthropological Institute of Great Britain and Ireland, reprinted by permission of Royal Anthropological Institute of Great Britain and Ireland.

Excerpts from "Transformation of Buddhism in China" by Chan Wing-tsit in *Philosophy East and West*, vol. 7, No. 3-4 (Oct. 1957-Jan. 1958), Copyright© 1957 by *Philosophy East and West*, reprinted by permission of *Philosophy East and West*.

Excerpts from *Understanding Media: The Extensions of Man* by Marshall McLuhan, Copyright© 1965 by McGraw-Hill Publishing Co., reprinted by permission of McGraw-Hill Publishing Co.

Grateful acknowledgment is made to the following publishers:

To Encyclopedia Britannica, Inc. for permission to quote from John Locke, "An Essay Concerning Human Understanding" in *Great Books of the Western World*, Vol 35 (BK. 11-chap. 33-5), Copyright© 1952 by Encyclopedia Britannica, Inc.

To Grove and Weidenfeld Publishing for permission to quote from *Happy Days* by Samuel Beckett, Copyright© 1961 by Grove Press; and *Watt* by Samuel Beckett, Copyright© 1959 by Grove Press.

To Oxford University Press for permission to quote from *The Sense of an Ending* by Frank Kermode, Copyright© 1967 by Oxford University Press; *The Poetry of T'ao Ch'ien* by James Robert Hightower, Copyright© 1970 by Oxford University Press.

To Princeton University Press for permission to quote from *Ancient Myth in Modern Poetry* by Lillian Feder, Copyright© 1971 by Princeton University Press; *A Source Book in Chinese Philosophy* by Chan Wing-tsit, Copyright© 1972 by Princeton University Press; *The I Ching: Book of Changes* by Richard Wilhelm, trans. by Cary Baynes, Copyright© 1967 by Princeton University Press; *The Literary Impact of the Golden Bough* by John Vickery, Copyright© 1973 by Princeton University Press.

To Thames and Hudson, Ltd. for permission to quote from *A Reader's Guide to William Butler Yeats* by John Unterecker, Copyright© 1959 by Noonday Press; *Myth and Ritual in Christianity* by Alan W. Watts, Copyright© 1954 by Thames & Hudson, Ltd.

To University of California Press for permission to quote from *The Rise of the Novel* by Ian Watt, Copyright© 1967 by University of California Press; *Samuel Beckett: A Critical Study* by Hugh Kenner, Copyright© 1968 by University of

viii *Acknowledgments*

California Press; *Time in Literature* by Hans Meyerhoff, Copyright© 1968 by University of California Press.

I should also like to express my gratitude to authors and publishers of the following works I have quoted:

To Charles Chen, ed., *Neo-Confucianism Etc.: Essays by Wing-tsit Chan*, Copyright© 1969 by Oriental Society.
To Leonard Cohen, *Beautiful Losers*, Copyright© 1967 by The Viking Press.
To K. A. Dhiegh, *The Eleventh Wing*, Dell Publishing Co., 1973, Copyright© by Nash Publishing Corporation.
To Mircea Eliade, *Cosmos and History: The Myth of the Eternal Return*, Copyright© 1959 by Harper Torch Books.
To Richard Ellmann, *Yeats: The Man and the Masks*, Copyright© 1948 by E. P. Dutton.
To Martin Esslin, *Samuel Beckett: A Collection of Critical Views*, Prentice-Hall Inc., 1965, Copyright© by Martin Esslin.
To Northrop Frye, *A Study of English Romanticism*, Copyright© 1968 by Random House, Inc.
To David Hesla, *The Shape of Chaos*, Copyright© 1971 by The University of Minnesota Press.
To Theodore Hunt, *Literature: Its Principles and Problems*, Copyright© 1906 by Funk & Wagnall Co.
To James Joyce, *Finnegans Wake*, Copyright© 1939 by The Viking Press.
To Arthur Koestler, *The Act of Creation*, Copyright© 1969 by Pan Books Ltd.
To Harry Levin, *James Joyce: A Critical Introduction*, Copyright© 1941 by New Directions Publishing Co.
To Frank L. Lucas, *The Drama of Chekhov, Synge, Yeats and Pirandello*, Copyright© 1963 by Cassell Co.
To Lu Hsün, "The Biography of Ah Q" in *Selected Works of Lu Hsün*, Copyright© 1983 by Jen-min ch'u-pan-she.
To A. A. Mendilow, *Time and the Novel*, Copyright© 1972 by Humanities Press.
To Morse Peckham, *Man's Rage for Chaos*, Schoken Books, 1967 Copyright© by Chilton Books Co.
To Henry Murray, "The Possible Nature of a Mythology to Come" in *Myth and Mythmaking*, Copyright© 1960 by George Braziller.
To William E. Soothill, *The Hall of Light*, Copyright© 1951 by Lutterworth Press.
To Sharon Spenser, *Space, Time and Structure in Modern Novel*, Copyright© 1971 by New York University Press.
To John Russel Taylor, *Anger and After: A Guide to the New British Drama*, Methuen Co., (London), 1963; U.S.A. Copyright© by Hill and Wang Inc.
To Hellmut Wilhelm, *Change: Eight Lectures on the I Ching*, Copyright© 1960 by Harper & Row Publishers.
To Laurence Sterne, *Tristram Shandy*, Copyright© 1940 by The Odyssey Press.
To Stark Young, trans., "The Seagull" in *Best Plays by Chekhov*, Copyright© 1956 by Random House, Inc.

CONTENTS

Preface   xiii
Abbreviations   xv
List of Illustrations   xvii

I. INTRODUCTION: CYCLIC MYTHS AS TEMPORAL SCHEMATA OF THE UNITY OF MAN AND THE COSMOS   1

II. ESSAYS ON CHINESE TEMPORAL SCHEMATA IN MYTH, RITUAL, PHILOSOPHY, POETRY, DRAMA, AND THE NOVEL
  A. ASPECTS OF CYCLIC MYTH IN CHINESE MYTHOLOGY
    1. Reconstruction of Fifteen Chinese Myths   9
      a. The Myth of the Creation   9
      b. The Myth of the Creation of Man   11
      c. The Myth of the Great Deluge: The Destruction of the Universe   13
      d. The Myth of Paradise Rebuilt and the Golden Age   13
      e. The Myth of the Silver Age   14
      f. The Myth of Passageways to Heaven   14
      g. The Myth of the Divine Conference and the Rebellion of Titan Ch'ih Yu   16
      h. The Myth of the Estrangement of Earth from Heaven   18
      i. The Myth of Disasters on Earth After the Estrangement   19
      j. The Myth of the God of Time and the Journey of the Sun   19
      k. The Myth of Titan P'eng Tsu's Sorrow Over the Ephemerality of Life   21
      l. The Myth of Titan K'ua Fu, Who Chased the Sun   22
      m. The Myth of Archer God Hou I, Who Shot Down Nine Suns   23
      n. The Myth of the Flight to the Moon   24
      o. The Myth of the Divine Administration   26

    2. The Temporal Envisagement in Chinese Mythology   32
      a. The Myth of the Estrangement of Earth from Heaven: A Myth of the Rise of Time and Mortality   32
      b. The Myths of Hou I and Titan K'ua Fu: Myths of the Rebellion Against Time   35
      c. The Myth of the Flight to the Moon: A Myth of the Reconciliation with Time   36
      d. The Myth of the Divine Administration: A Myth of the Eternal Becoming   37

B. ASPECTS OF CYCLIC MYTH IN CHINESE RITUALS AND EARLY BELIEFS

  1. Ritual: A Mode of Actualization 40
  2. Totemism 40
  3. The Sacrifice to Heaven and Earth 41
  4. The Sacrifice to the Four Directions 43
  5. The Sacrifice to the Cosmic Mountain 44
  6. The Institution of the Hall of Light 44
  7. The Ancestor Worship Under a Personal Heaven 48
  8. The Mandate of an Impersonal Heaven 49
  9. Tao as a Totalistic Concept of an Ideal and Universal Order 49
  10. Summary: From Peasant Omen to a Microcosm 50

C. ASPECTS OF CYCLIC MYTH IN CHINESE PHILOSOPHY

  1. Philosophy: A Mode of Rationalization 60
  2. Yin-yang School: Dual-multiple Cosmogony and Penta-cyclical Cosmology 60
  3. Confucianism: The Unity of Man with Cosmos 62
  4. Taoism: The Harmony of Man with Nature 66
  5. Chinese Buddhism: The Identification of Man with the Universal Mind 69
  6. Summary: Cyclic Myth in the Core of Chinese Culture 72

D. ASPECTS OF CYCLIC MYTH IN CHINESE LITERATURE

  1. The Mythic and Philosophical Aspects of the Poems of "Encountering Sorrow" and "The Return" 78
     a. "Encountering Sorrow" (Li sao): A Mythic Quest for the Unity with Cosmos 79
     b. "The Return" (Kuei ch'ü lai tz'u): A Pastoral Quest for Harmony with Nature 86

  2. The Mythic and Philosophical Aspects of the Plays of *The Peony Pavilion* and *A Dream in Han-tan Inn* 90
     a. The Metaphysical Basis of T'ang Hsien-tsu's Plays 90
     b. *The Peony Pavilion* (Mu-tan t'ing): An Incessant Pursuit from Limbo to Resurrection 92
     c. *A Dream in Han-tan Inn* (Han-tan meng chi): An Ordeal Descent into Eternity 98

  3. The Mythic and Philosophical Aspects of the Novels of *The Journey to the West* and *The Dream of the Red Chamber* 102
     a. The Historical Background 102
     b. *The Journey to the West* (Hsi-yu chi): A Celestial Progress to Divinity 103
     c. *The Dream of the Red Chamber* (Hung-lou meng): A Terrestrial Journey to Disillusionment 111

4. Summary: the Cyclic Schema in Chinese Poetry, Drama, and the Novel   *121*
   a. Archetypal Images   *121*
   b. Symbolic Phases   *124*

III. ESSAYS ON WESTERN TEMPORAL SCHEMATA IN MYTH, RITUAL, PHILOSOPHY, POETRY, DRAMA, AND THE NOVEL

A. THE SURVIVAL OF CYCLIC MYTH IN WESTERN LITERATURE

1. Historical Background: Linear Eschatology Sustained Through Cyclic Symbolism   *127*
2. The Survival of the Cyclic Myth in Natural Symbolism   *130*
3. Dante's *Divine Comedy*: A Spiral Transcendence to the Centre of Circles   *131*
4. Milton's *Paradise Lost*: A Fall into Linearity   *138*
5. Goethe's *Faust*: A Striving for Meaningful Duration   *141*
6. Shelley's "Ode to the West Wind": A Poet's Vision at the Cycle's Centre   *144*
7. Yeats' *A Vision*: A Prophecy about Cyclic Cosmos   *148*
8. Joyce's *Finnegans Wake*: A Viconian Myth of the Course of Humanity   *149*

B. ASPECTS OF TIME IN EARLY ENGLISH NOVELS: *MOLL FLANDERS*, *PAMELA*, AND *TRISTRAM SHANDY*

1. The Philosophical Background   *152*
2. Defoe's *Moll Flanders*: Memoirs of Sensory Details of Successive Moments   *156*
3. Richardson's *Pamela*: Accounts of Sentimental Moments with Intense Feeling   *158*
4. Sterne's *Tristram Shandy*: Portrayals of the Associative and the Sensational   *159*

C. A CASE STUDY OF THE VISUAL IMPERATIVE IN *WATT* AND *BEAUTIFUL LOSERS*

1. The Philosophical Background   *163*
2. Beckett's *Watt*: The Visual Imperative as an Index of the Nature of Language   *166*
3. Cohen's *Beautiful Losers*: The Visual Imperative as an Index of the Nature of Consciousness   *169*

D. SUMMARY: THE MYTHOPOEIC IMPOTENCE OF THE MODERN WESTERN MAN   *177*

IV. THE CULTURAL CONVERGENCE AND THE AGE'S "TIME SPIRIT"

A. A CASE STUDY OF THE PHILOSOPHICAL MESSAGE IN THE WORKS OF IBSEN, CHEKHOV, LU HSÜN, BECKETT, AND PINTER

1. The Cultural Convergence and the Age's "Time Spirit" — *179*
2. Ibsen's *The Wild Duck* — *181*
3. Chekhov's *The Seagull* — *183*
4. Lu Hsün's *The Biography of Ah Q* — *185*
5. Beckett's *Happy Days* — *187*
6. Pinter's *The Homecoming* — *190*

V. CONCLUSION: CYCLIC MYTH AS INDEX OF CULTURE

A. CYCLIC ONTOLOGY VERSUS LINEAR ESCHATOLOGY — *193*

B. CYCLIC MYTH AS AN INFORMING STRUCTURE OF LITERARY WORKS AND AN INDEX OF CULTURE — *196*

BIBLIOGRAPHY — *199*

INDEX — *209*

PREFACE

I first became interested in this subject years ago when I was producing a TV series on Chinese culture. It took a few years for the initial idea to take its current shape. I would like to take this opportunity to thank Professor Graham Good of the English Department and Professor Florence Chia-ying Yeh-Chao of the Department of Asian Studies of the University of British Columbia for devoting their profound scholarships to help me solve problems that often baffled me in the first draft. I am also indebted to Professor Jan Walls of Simon Fraser University and Professor Marketa Goetz-Stankiewicz of the Department of Germanic Studies of UBC, who made critical readings of one part or another of my first draft and offered valuable criticism.

I would like to express my gratitude to those scholars whose critical works or translations have been most frequently consulted: Northrop Frye on myth and literature; Mircea Eliade on cosmos and history; Georges Poulet, Hans Meyerhoff and A. A. Mendilow on time; Joseph Needham, Nathan Sivin and Fredirick Mote on Chinese concept of time; Joseph Campbell, Robert Harrison, Harry Slochower, Philip Wheelwright, John Vickery and Lillian Feder on myth and literature; Susanne Langer, Chen Wing-tsit, Fang Thomé, Fung Yu-lan, Amaury de Riencourt, Frederick Mote, Richard Wilhelm, Hellmut Wilhelm and K. A. Dhiegh on philosophy; Chang Kuang-chih, Yüan Ke, Shen Yen-ping, Ting Ying, Lao Kan and Tu Er-wei on Chinese mythology; William Soothill, Clyde Kluckholm, Huang Wen-shan and Liu Chieh on ritual; Erich Auerbach, Georges Poulet, and Thomas Bergin on Dante; Ian Watt on English novel; Harold Bloom and Richard Ellmann on Yeats; Harry Levin and W. Y. Tindall on Joyce; David Hesla and Hugh Kenner on Beckett; Sandra Djwa on Leonard Cohen; David Hawkes, James Robert Hightower and Lucian Miller on Chinese poetry; Cyril Birch and Edward Gunn on Chinese drama; C. T. Hsia, Glen Dudbridge, and James Fu on the Chinese novel. I am very grateful to C. T. Hsia and Joseph Campbell, whose critical writings have been most inspirational.

I wish to thank Miss Katie Eliot and Mr. Robin Baker for their verbal comments and intellectual stimulation. My special thank goes to Connie Chen for her personal support during the difficult period in this project. Finally, to Lissa, Christina and Catharine, without whose constant encouragement and support this book could not have been completed, I dedicate it with love.

R. S. C.
January, 1992.
Vancouver, B.C.

Abbreviations of Titles of Source Books of Chinese Mythology

CLMS	Chou li: Mei shih (Matchmaker: *Rites of Chou*, c. 100 B.C.)
CSCMP	Chou shu: Ch'ang mai p'ian (Ch'ang mai: *Documents of Chou*, c. 300 B.C.)
CTCN	Ch'u tz'u: Chao hun (Summon the Soul: *The Songs of Ch'u*, 342-290 B.C.)
CTSKT	Chuang tzu: Keng sang ch'u (Keng sang ch'u: *Chuang tzu*, 399-295 B.C.)
CTTW	Ch'u tz'u: T'ien wen (Heavenly Questions: *The Songs of Ch'u*, 342-290 B.C.)
FSTI	*Feng-su t'ung-i (General Introduction to Customs*, c. A.D. 150)
HFTSK	Han fei tzu: Shih kuo (Ten Faults: *Han fei tzu*, c. 233 B.C.)
HFTSL	Han fei tzu: Shuo lin (Collected Persuasions: *Han fei tzu*, c. 233 B.C.)
HNTCS	Huai nan tzu: Ching shen (Spirituality: *Huai nan tzu*, 178-122 B.C.)
HNTHW	Huai nan tzu: Hsiu wu (Administration: *Huai nan tzu*, 178-122 B.C.)
HNTLM	Huai nan tzu: Lan meng (On Antiquity: *Huai nan tzu*, 178-122 B.C.)
HNTPC	Huai nan tzu: Pen ching (Major Treaties: *Huai nan tzu*, 178-122 B.C.)
HNTSL	Huai nan tzu: Shuo lin (Collected Persuasions: *Huai nan tzu*, 178-122 B.C.)
HNTST	Huai nan tzu: Shih tse (Observances of Seasons: *Huai nan tzu*, 178-122 B.C.)
HNTTH	Huai nan tzu: Ti hsing (Geography: *Huai nan tzu*, 178-122 B.C.)
HNTTW	Huai nan tzu: T'ien wen (Astronomy: *Huai nan tzu*, 178-122 B.C.)
HTYNCF	Huang ti: Yüan-nü chang-fa (Strategy of Goddess of Mystery: *Yellow Emperor*)
IS	*I shih (The Interpretative History*, A.D. 255)
KPWC	*Kuang po wu chih (Great Record of the Marvellous*, A.D. 1607)
KT	*Kui tsang (Reverting Deposit*, c.1766-1123 B.C.)
KYCY	Kuo yü: Ch'u yü (The Ch'u State: *Remarks Concerning the States*, c. 400 B.C.)
KYC'Y	Kuo yü: Chin yü (The Chin State: *Remarks Concerning the States*, c. 400 B.C.)
LCYL	Li chi: Yüeh ling (Monthly Observances: *Book of Rites*, c. 100 B.C.)
LSCCMCC	Lü-shih ch'un-ch'iu: Meng-ch'un chi (Spring: *Lü-shih ch'un-ch'iu*, c. 235 B.C.)
LSCCYSP	Lü-shih ch'un-ch'iu: Yu shih p'ian (Genesis: *Lü-shih ch'un-ch'iu*, c. 235 B.C.)
LSHCSCYC	Lu shih hou chi: Ch'ih yu chuan
LTHTP	Lieh tzu: Huang Ti p'ian (Yellow Emperor: *Lieh tzu*, c. 314 B.C.)
LTTW	Lieh tzu: T'ang wen (Questions of T'ang: *Lieh tzu*, c. 314 B.C.)
LYHT	*Lung-yü he-t'u (Pictures Carried by Dragon-fish From the River)*
MTFK	Mo tzu: Fei kung (Condemnation of War: *Mo tzu*, 501-416 B.C.)
MTTC	*Mu T'ien Tzu chuan (The Travels of Emperor Mu*, 408 B.C.)
PHTWH	Pai hu t'ung: Wu hsing (Five Elements: *Collection of Five Classics by Confucius in White Tiger Hall*, A.D. 32-92)
PT	*Pu tz'u (Inscription on Oracle Bones*, 1300 B.C.)
SCLH	Shu ching: Lü hsing (The Prince of Lü Upon Punishment: *Book of Documents*, c. 300 B.C.)
SCSCPSHPC	Shih chi Ssu-ma Chen pu san-huang pen chi (*Ssu-ma Chen's Amendments to Chronicles of Three Emperors in Records of History*, 145-86 B.C.)
SCTYM	Shu ching: Ta Yü mo (Counsels of Great Yü: *Book of Documents*, c. 300 B.C.)

Abbreviations

SCWTPC	*Shih chi*: Wu-ti pen chi (Chronicles of Five Legendary Kings: *Records of History*, 145–86 B.C.)
SHC	*Shen-hsien chuan (Biographical Notes of Immortals*, A.D. 250–330)
SHC,TT,PC	*Shan-hai ching, t'u-tzan, pu-chu (Annotations and Commentaries to Pictures of the Mountain-sea Classic*, A.D. 276–324)
SIC	*Shu-i chi (Collection of Notes on the Wonderful*, A.D. 550)
SPSHP	*Shih-pen shih-hsing p'ian (On Surnames)*
SWCT	*Shuo-wen chieh-tze (The Etymological Dictionary*, A.D. 100)
SWLC	*San-wu li chi (Annals of Three Emperors and Five Kings*, c. A.D. 220-280)
TPYL	*T'ai-ping yü-lan (Encyclopaedia Royal T'ai-ping*, A.D. 983)
TWSCCC	*Ti-wang shih-chi chiao (Annotation Annals of Kings and Princes)*
TYTT	Tu Yü's *T'ung tien (Comprehensive Historical Records*, A.D. 735–812)
WCSIC	Wang Chia's *Shih-i chi (Records of Matters Omitted in the Annals of the Empire*, c. A.D. 390)
WYLNC	*Wu-yün li-nian chi (Annals of Five Legendary Kings)*

LIST OF ILLUSTRATIONS

1. Sketch of Supernal Goddess Nü Wa and Sovereign God Fu Hsi   12
2. Sketch of the Ten Suns on the Cosmic Tree   21
3. Sketch of the Archer God Shooting the Ten Suns   24
4. Sketch of the Flight to the Moon   26
5. Chart of the Hierophany of Gods of Directions   30–31
6. Diagram of the Institution of Ming T'ang, the Hall of Light   46
7. Diagram of the Primal Arrangement of Pa Kua in Pairs of Antitheses   52
8. Chart of the Primal Arrangement of Pa Kua   54
9. Chart of the Evolution of Ideas in Chinese Culture   58–59
10. Diagram of Dante's Conceptions of the Cosmos   133
11. Diagram of Dante's Conceptions of the Hell   134

A COMPARATIVE STUDY OF CHINESE AND WESTERN CYCLIC MYTHS

Essays on Temporal Schemata in
Myth, Ritual, Philosophy, and Literature

I. INTRODUCTION: CYCLIC MYTHS AS TEMPORAL SCHEMATA OF THE UNITY OF MAN AND THE COSMOS

This comparative study of selected Chinese and Western literary works is a collection of essays on temporal schemata in myth, ritual, philosophy, poetry, drama, and the novel. Time is the essential experience of human existence. It is usually comprehended as a linear progression of succession which is never experienced as a whole; as soon as the future becomes the present, the present becomes the past. The fact that things exist through time demands a second concept, that of permanence or duration. As an adequate description of time, duration must be understood as arising from the successive flux of time, yet conversely, only within the background of duration is the emergence and the human awareness of succession possible. Without duration, time is the mere succession of heterogeneous moments; while without succession, duration is but an all-inclusive and unchanging present. Duration then becomes the sustaining quality of time and the actual "co-existence" of past, present, and future. The concept of duration has important ramifications for consciousness; the individual as a self-centred form of life must endure by opposing its permanency to its transitoriness in time. And the highest state of consciousness, whether transcendental or mythical, is the awareness of the "co-existence" of past, present, and future, or the experience of time as a whole.

Myth, as Mircea Eliade perceives it, is a "complex system of coherent affirmation about the ultimate reality of things,"[1] a system which constitutes a metaphysical validation of human existence, or a primitive ontology that asserts an intimate relationship between man and the cosmos. In its sociopoeic aspect, myth is a dramatization and a rationalization of the fundamental needs of the society. Here Bronislaw Malinowski's definition of myth as "a narrative resurrection of a primeval reality, told in satisfaction of deep religious wants, moral cravings, social submissions, assertions, even practical requirements"[2] serves as the best description of its vital function in primitive society. Finally in its psychopoeic aspect we turn to the pioneering work of Carl Jung where "myths are original revelations of the preconscious psyche, involuntary statements about unconscious psychic happenings."[3] Here myth is an expression of instinctual drives, repressed wishes, fears and conflicts, and remnants of the

---

[1] Mircea Eliade, *Cosmos and History* (New York and Evanston: Harper Torchbooks, 1959), p. 3.

[2] Bronislaw Malinowski, "Myth in Primitive Psychology" in *Magic, Science, and Religion, and Other Essays* (Boston, 1948), pp. 78-79; cited by Lillian Feder in *Ancient Myth in Modern Poetry* (Princeton, New Jersey: Princeton University Press, 1971), p. 6.

[3] C. G. Jung and C. Kerényi, *Essays on a Science of Mythology*, trans. from the German by R. F. C. Hull, rev. ed. (New York, 1963), p. 79; cited by Feder, *Ancient Myth in Modern Poetry*, p. 50.

collective unconscious. Whether in its mensopoeic, sociopoeic, or psychopoeic aspects (and all are involved in apprehending the cyclic myth,) myth is the original touchstone of culture. We will maintain that myth circumscribes a verbal universe in which man's psychological, social, religious, and philosophical lives are really one in the juxtaposition and succession of the stages of gods, heroes, and mortals.

In this analysis, with functional similarity in regard to soothing the psyche, stabilizing the society, and promoting cultural continuity, we will not enter into the controversy surrounding the primacy of myth or ritual. Although myth connotes a way of envisaging, and ritual, as a temporal sequence of acts, denotes a way of doing, their functional relation is both inevitable and interdependent. Lillian Feder best renders this mutual dependence as: "Myth clarified the prescribed action of rites, and rites enacted mythical narrative in stylized dramatic form."[4] We would thus agree with Northrop Frye that myth should be understood as "the central informing power that gives archetypal significance to the ritual," while a ritual may in turn suggest the origin of the narrative of myth.[5]

Aside from ritual, myth tends to reveal a body of metaphysical thought whose development may be consequent to the original myth. This process was suggested by Susanne Langer: "When the mythical mode is exhausted, natural religion is superseded by a discursive and more literal form of thought, namely philosophy,"[6] and amplified by Northrop Frye to include literature "which inherits the fictional and metaphorical patterns that identify aspects of human personality with the natural environment."[7] In a sense, then, both philosophy and literature merge inseparably with myth, while literature especially inherits plots and characters, and themes and images which are complications (in Northrop Frye's terms, "displacements") of similar elements in myth. It is this universalizing tendency of myth which justifies our intention to investigate and adumbrate the temporal schemata engendered to cope with the disparity between human and cosmic time.

To compare the cultures of East and West, especially to compare the literatures of China and Western Europe, one is inevitably doomed to uncertainty, whether the study of comparative literature is regarded as "the comparison of literature with other spheres of human expression" or as "the study

---

[4] Feder, *Ancient Myth in Modern Poetry*, p. 5.

[5] Northrop Frye, *Fables of Identity* (New York: Harcourt, Brace and World, 1963), p. 15.

[6] Susanne Langer, *Philosophy in a New Key* (New York: A Mentor Book, 1951), p. 172.

[7] Northrop Frye, *A Study of English Romanticism* (New York: Random House, 1968), pp. 4–5.

of literature beyond the confines of one particular country."[8] The abysmal remoteness of the two cultures through history tends to magnify any perceived correspondence to the point of assured conclusions where only rough similarities may, in fact, exist. In the comparison of literature with other areas of knowledge the very homogeneity of Chinese culture poses a problem. The integration of philosophy and literature, the coherence of religion and philosophy, the correlation of philosophy and politics, the similarities between various philosophic schools, the aesthetic coalescence of lyric verse and music, and painting and poetry; these concerns present a united front, almost a barrier to the inquisitive Western mind.

In the above perspective, a comparative study of literary themes, especially of archetypal image and pattern, that which is most fundamental and beyond temporal and spatial barriers is likely to be the most urgent and constructive task for the student comparing Chinese and Western literature. As the ultimate expression of psychic truth and the primordial yet autonomous form of the human spirit, the cyclic myth naturally arouses one's attention.

As one of the most vital archetypes of human consciousness, the cyclic myth is inevitably temporally centred. It is a myth of eternal return, the myth of periodic becoming and perpetual regeneration. Its themes represent man's vision and manipulation of the unfathomable cosmic setting of his existence, especially the awful consciousness of his biologically irreversible death against the rebirth of the sun, moon, and the seasons. This linear consciousness of the self against nature would remain an imponderable nightmare were it not for man's mythmaking imagination which recognizes in nature the permanent symbol and divine rule, and assimilates himself to it to define his life and being. In essence, the cyclic myth identifies man with the periodic becoming and regeneration in nature, and guarantees personal duration against the flux of time.

In the mythopoeic perspective, nature is cyclical; there is everywhere a concept of the end and beginning of a temporal period, and there are always instant repetitions of images and obvious metaphors ready to convey a dawning concept of being. In heavenly motion, solar alternation, lunar regularity, and seasonal rotation, there appears a pattern of cyclic significance which offers an optimistic view of life in general. Northrop Frye best summarizes this as: "Myth seizes on the fundamental element of design offered by nature—the cycle, as we have it daily in the sun and yearly in the seasons—and assimilates it to the human cycle of life, death, and rebirth."[9]

---

[8] Henry H. Remak, "Comparative Literature: Its Definition and Function," in Newton P. Stallknecht and Horst Frenz, ed., *Comparative Literature: Method and Perspective* (Carbondale and Edwardsville: Southern Illinois University Press, 1971), p. 1.

[9] Frye, *Fables of Identity*, p. 32.

Above all, the identification of human and natural cycles presents a conviction of the periodic regeneration of time which in turn is a revolt against the historical or linear time of succession. Here we rely greatly on the research of Mircea Eliade, especially his *Cosmos and History*. According to Mircea Eliade, the cyclic ontology makes the abolition of historical linear time possible through a reduction of events to categories, individuals to archetypes, space to the "centre," and time to the original time of creation. Then, through the repetition of meaningful paradigmatic acts or archetypal gestures, the linear time of succession is suspended as man participates in the time of pure duration, the mythical time of infinite perpetuity.[10] In this regard the cyclic myth can be seen as an identification of man with the cyclical structure of nature from which he derives a cyclic notion of time that makes his life meaningful and his being harmonious with the cosmic setting of incessant regeneration which surrounds him.

Thus the first task of this study is to trace the origins, formation, and presentation of that most pervasive mentality or schema known for ages in Chinese culture as **"T'ien Jen Ho I,"** literally the unity of man and cosmos; and further, to engender insight into the cyclic myth as a whole by contrast and comparison to its Western counterpart. The first step is a simplification of terminology; the schema of "T'ien-jen ho-i" is hereafter to be rendered as "the Chinese cyclic myth" which immediately suggests the common importance of cyclicity to both cultures while allowing for more inclusive application in the relevant areas of study.

In order to bring the Chinese cyclic myth into a coherent perspective, efforts are devoted to the construction and analysis of fifteen Chinese myths which concern temporal consciousness. The temporal schema is then traced through increasing levels of abstraction: ritual, natural religion, and finally philosophical speculation, for each has been incorporated within the cyclic mentality of the homogeneous Chinese culture. We will trace the formation of the cyclic mentality in archaic Chinese mythology by investigating its narrative origin in the dramatic enactments of natural religion in totemism, manaism, and ancestral worship, in the Chinese rituals of sacrifices to Heaven, Earth, and the Cosmic Mountain, and in the institutions of Monthly Observances and the Hall of Light. Then we will investigate ontological assertions of the unity or harmony of man and the cosmos in major philosophical schools such as the Yin-yang School, Confucianism, Taoism, and Chinese Buddhism. With the abstraction of the Chinese cyclic myth broadly defined and its existence in the core of Chinese culture firmly assured, we turn to elaborate its presentation in Chinese literature. Here, we are indebted to the works of Northrop Frye and Mircea Eliade who, despite the limited inclusiveness of their theories, have provided the framework

---

[10] Cf. Eliade, *Cosmos and History*, pp. 35-36.

for studying the relationship between myth and literature. In literature the cyclic myth is most evident in thematic structure and symbolism which depicts the cyclic cosmogony within which the action of the work occurs, and in the pattern of plot usually described as the cyclic "quest of the hero." This latter pattern has been a frequent subject of literary theorists, receiving its broadest treatment in Northrop Frye's *Anatomy of Criticism*, but is also ably represented in Robert Harrison's essay, "Symbolism of the Cyclical Myth in 'Endymion'."[11]

In Chinese literature, the cyclic myth is self-evident in the quest for the unity with the cosmos in Ch'ü Yüan's "Encountering Sorrow" (Li sao). In his "amour quest" on a cosmic journey of circuitous progress, the poet is assured a harmony with the cosmos and granted a re-ascendence to the hub of the universe, a cosmological reconciliation. The cyclic myth is also evident in T'ao Ch'ien's "Rhapsody on the Return" (Kuei ch'ü lai tz'u). The poet's homecoming to the pastoral existence is a spiritual quest for harmony with nature, for only in nature, as a state of spontaneity in an incessant flux of transformation, can man unite himself with the infinite and enjoy the regeneration of eternal return.

In Chinese drama, the cyclic myth is most apparent with T'ang Hsien-tsu's dream plays. T'ang is a student of Lo Ju-fang, a prominent philosopher of Neo-Confucian thought, who regarded the perpetual renewal of life as a ceaseless vitality and the animating principle of the universe, equated human nature with the vitality of life, and assimilated the self to the regeneration of the universe through humanity as an incessant power of creation. This vital strain of Lo's thought is especially explicit in *The Peony Pavilion (Mu-tan t'ing)* and *A Dream in Han-tan Inn (Han-tan meng chi)*. *The Peony Pavilion*, elaborating an incessant pursuit from limbo to resurrection, is a dramatic romance underscored by the cyclic myth with the affectionate self at the thematic centre. It is a revelation of man's unity with the perpetual renewal of the universe through the triumph of love over life and death. With *A Dream in Han-tan Inn*, we descend to a dramatic structure analogous to C. G. Jung's archetype of transfiguration and reconciliation. It is a play of enlightenment through an apocalyptic dream, an ordeal descent into eternity. T'ang believed that the harmonious unity of self and cosmos relies on the attainment of pure being through true selfhood accessible through awakening from the apocalyptic dream.

In the Chinese novel, the cyclic myth is most explicit in Wu Ch'eng-en's *The Journey to the West (Hsi-yu chi)* and Ts'ao Hsüeh-ch'in's *The Dream of the Red Chamber (Hung-lou meng)*. *The Journey to the West* is a Rabelaisian mockery in the mythic mode. It is a divine comedy of quest, progress, and reconciliation of the terrestrial, the demonic, and the underworld with the

---

11 Robert Harrison, "Symbolism of the Cyclic Myth in Endymion," in John B. Vickery, ed., *Myth and Literature: Contemporary Theory and Practice* (Lincoln: University of Nebraska Press, 1966), p. 230.

infinite paradise of Buddha and the heaven of celestial deities. As a quest for divinity and eternal existence, the mythic implication of *The Journey to the West* is similar to the cyclic significance in the rituals of Sacrifices to Heaven, Earth, and the Cosmic Mountain, or in the institution of the Hall of Light (Ming t'ang). With *The Dream of the Red Chamber*, we encounter a structure which recalls Robert Harrison's "quest" progression. The cyclic myth is most evident in the novel's initial calls for the celestial spirits and earth-bound mortals to embark on a quest, the subsequent descents to the mundane world or the dream world of the Garden of Great Wonder (a journey of ordeal and disillusionment through a transient life for the questing souls, and a terrestrial life of sense and sentiment for the descending spirits), and the "fulfilment" of the quest, which brings a reconciliation and return for the descending spirits, and a spiritual ascent for the mortals. Thus we are able to summarize essential archetypal images and symbolic phases to depict the homogeneous cyclic schemata present in every sphere of human experience in Chinese culture.

Furthermore, a brief study of the transmutation of the cyclic myth in Judaeo-Christian culture and a subsequent survey of the survival of the cyclic myth in Western European literature, despite the dominant eschatology of Christianity, serve as the focus of comparison between East and West. The survival of the cyclic myth in natural symbolism in Western literature is adumbrated through Dante Alighieri's spiral transcendence to the centre of circles in *Divine Commedia* and John Milton's mythopoeic description of the fall to linearity in *Paradise Lost*. It is explicated in Johann Wolfgang von Goethe's upward striving for meaningful duration in *Faust*, and Percy Bysshe Shelley's vision at the cycle's centre in "Ode to the West Wind." It is evident in William Butler Yeats' prophecy about cyclic cosmos in *A Vision* and James Joyce's Viconian myth of the course of humanity in *Finnegans Wake*; both display awareness of modern man's inability to embrace a mythopoeic perspective. They attempt to rehabilitate the cyclic myth for modern man and restore to art its cosmogonic function of explaining the true nature of reality.

The shifts in temporal thought are further traced from the Renaissance to the eighteenth century, and their importance in the rise of the English novel is subsequently investigated. The discussion focuses on Daniel Defoe's *Moll Flanders*, Samuel Richardson's *Pamela*, and Laurence Sterne's *Tristram Shandy*; each novel's formal realism dictated in part by the prevailing attitudes toward time. The eighteenth-century writer experienced life as a successive series of moments connected by a duration based on personal memory and multiplicity of sensation. Defoe succeeded in providing sensory details of the moment but failed to describe the sentimental aspect of the moment. Richardson succeeded in portraying intense feeling but over-emphasized the sentimental side which resulted in an issue of paradoxical morality. Sterne transcended the moral issues by substituting plot, sequence, and causality with a structure based on the operative character of consciousness; the associative and the sensational aspects account for the external and internal structures of the novel. The progress of the novel is multidimensional; the apparent disorder of the structure is actually order organized according to the rules of the Lockean mind. As the successive nature of language confines it to organizing the flux of consciousness into linear "bits" of information, Sterne seriously recognized the temporal limitation of language. This awareness and his definitive portrayal of time delineated a model which Samuel Beckett and Leonard Cohen would begin to duplicate one hundred and fifty years later.

Since Ludwig Wittgenstein has exposed the failure of language as a means of apprehending reality, the modern novelist faces a universe which has no meaning except for the words he uses. This universe is private since no one can experience another's experience. To surpass this privacy, the novelist assumes a common process of perception among his readers, and exploits the experience of words as objects, and the narration as a surface of the visual imperative.

Thus in *Watt*, a novel of language as futile depiction, Samuel Beckett adopts the visual imperative primarily as an index of the nature of language. In each case it reinforces the falseness of verbal reality. It eventually sends the protagonist, a Cartesian hero, to an insane asylum and returns the reader to Beckett's abyss which separates the world from the word. However in *Beautiful Losers*, a novel of language as transcendable obstacle, Leonard Cohen adopts the visual imperative as an index of the nature of consciousness. In each case the visual imperative helps the reader to cross psychic barriers (especially the tyranny of language and history). Eventually, as the correspondence between the visual imperative and superverbal consciousness is maintained, a liberated consciousness (a state of sainthood) is obtained.

Ascending from Beckett's pessimistic abyss with Cohen's liberated consciousness, we have completed one Eastern and one Western circuit through myth, ritual, philosophy, poetry, drama, and the novel. These analyses of Chinese and Western temporal schemata show as many similarities as discrepancies. However, before comparing the origins and transformations of the cyclic myth in both cultures, it is imperative to explore the aspects of a philosophical approach to literature. Hence, case studies of the philosophical message in Henrik Ibsen's *The Wild Duck*, Anton Chekhov's *The Seagull*, Lu Hsün's *The Biography of Ah Q*, Samuel Beckett's *Happy Days*, and Harold Pinter's *The Homecoming* are presented. This reaffirms the supposition of the phenomenon of cultural convergence, the school of "Geistesgeschichte" (Intellectual History), which views literature as a parameter of each age's "time spirit," and finds that authors of different countries and epochs may perceive the problem independently, yet arrive at the same solution.

In conclusion, the essential cultural and literary distinctions between the Chinese and the Western worlds will be revealed as both cosmological and ontological. Thus by a general delineation of the cyclic myth in both cultures, the book offers an analysis of the cyclic schema as an informing structure of literary works or an index of culture, and yields a further understanding of the cyclic myth as a whole. Indeed, the uniqueness of the independence of Chinese literature and its remoteness to Western assumptions offer a double test. Arthur E. Kunst's conclusion on the ultimate objective of the comparative study of Asian and European literatures as "the creation of a truly comprehensive theory of literature... based on a knowledge of independently evolved imaginative traditions"[12] serves as the best starting point of this paper. While such an ambitious scope is yet beyond realistic expectation at this time, our comparison will hopefully reveal areas of further study as well as illuminate a concept central to both cultures.

---

[12] Arthur E. Kunst, "Literature of Asia," in Stallknecht, *Comparative Literature: Method and Perspective*, p. 323.

II. ESSAYS ON CHINESE TEMPORAL SCHEMATA IN MYTH, RITUAL, PHILOSOPHY, POETRY, DRAMA, AND THE NOVEL

A. ASPECTS OF CYCLIC MYTH IN CHINESE MYTHOLOGY

1. Reconstruction of Fifteen Chinese Myths

Any study of the earliest Chinese myths must acknowledge the insufficiencies of source documents. To minimize the vagaries of three thousand years of natural upheaval is to maximize the human factor, which in the case of the infamous Burning of Books of 220 B.C. was equally disastrous. Ignoring the more fantastic events of history, one still must deal with the well-meaning distortions of euhemerizing humanists, rationalizing naturalists, and fanatic religious sectarians. Fortunately for our purposes we are more concerned with the sustained influence of the overall fragmentary corpus, and the reappearance of earlier ideas in later irrefutable documents tends to confirm the accuracy of original models. Much remains to be done in reconstructing early Chinese mythology, and some of the problems may be resolved in future archaeological discoveries.

To avoid textual errors inherent in the fragmentary nature of the earliest source documents, we have confined ourselves to Chinese ancient classics. The Chart of Abbreviations of Titles of Source Books of Chinese Mythology provides bibliographical data for all sources used in rendering these fifteen myths. Although seven of the selected books date from after 100 A.D., they are either records of legends tacit in oral tradition, or books of etymology and encyclopedias of quotations from ancient lost texts. Furthermore, we limit source fragments to those which are identifiable with traces in classics and with legendary texts believed to have existed.

a. The Myth of the Creation

(1) In the beginning there was Chaos, egg-like and pregnant with P'an Ku, the Fathomless Antiquity.[1] This primal god lay in the dark womb of Chaos for nearly eighteen thousand years before his violent birth.[2] Bearing a dragon's head upon a serpent's body,[3] he awoke and fought the darkness with an axstone. P'an Ku became the medium of the forming world, growing between the clear and weightless matter which rose as the sky, and the dark and heavy

---

[1] Cf. Morohashi Tetsuji, *Dai kanwa jiten* (Tokyo: Daishukan shuten, 1968), p. 113.

[2] Adapted from SWLC, quoted by TPYL, II, cited by Yüan Ke, *Chung-kuo shang-ku shen-hua* (Peking: Chung-hua shu-chü, 1960), p. 37.

[3] Adapted from WYLNC, quoted by KPWC, IX, cited by Yüan Ke, p. 34. For the abbreviation, cf. Abbreviation of Titles of Source Books of Chinese Mythology.

which descended as the earth.[4] His breathing was the wind and his puffing the storm. The opening and closing of his eyes caused the turning of day and night.[5] When he was angry the weather was gloomy and when he was happy it was fine.[6]

This process continued for another eighteen thousand years, until the sky was extremely high and the earth tremendously thick.[7] Then the transformation began. P'an Ku's gigantic limbs and entrails became the four extremes of the earth and the five greatest mountains; his flesh turned into fertile soil, his hairy skin into grasses and trees; his sweat became rain and marshes, his veins and blood formed the rivers and streams; his breath was transformed into wind and cloud, his voice became the thunder, while his left eye became the sun and his right the moon; finally, his hair changed into myriads of stars and constellations.[8] Thus the universe was formed from the son of Chaos.[9]

(2) The God of the Central Region was called Chaos (Hun Tun)[10]—a primal simplicity in which myriad things were confounded and not yet separated from each other.[11] It was undefined and yet complete in itself.[12] It was an insubstantial form of a round,[13] bird-like shape with six feet and four wings but no openings on its face.[14] The God of the South Sea was called Ephemeral (Shu) and the God of the North Sea was called Swift (Hu). From time to time Ephemeral and Swift came together for a meeting in the territory of Chaos, and Chaos treated them very generously. In order to repay Chaos' kindness, Ephemeral and Swift decided to bore seven openings on his face. They bored one hole on him every day, but on the seventh day Chaos died.[15]

---

[4] Adapted from *SWLC*, quoted by *TPYL*, II, cited by Yüan Ke, p. 37.

[5] Adapted from *WYLNC*, cited by Ma Su, *I SHIH*, I (Taipei: Kuang-wen shu-chü, 1969, duplication).

[6] Adapted from *SIC*, cited by Yüan Ke, p. 38.

[7] Adapted from *SWLC*, quoted by *TPYL*, II, cited by Yüan Ke, p. 37.

[8] Adapted from *WYLNC*, cited by Ma Su, (I); also cf. Yüan Ke, p. 38.

[9] Cf. Shen Yen-ping, *Chung-kuo shen-hua yen-chiu* (Taipei: Hsin-lu shu-chü, 1929), p. 47; also cf. Yüan Ke, pp. 37–38.

[10] Cf. Burton Watson, *The Complete Works of Chuang Tzu* (New York and London: Columbia University Press, 1968), p. 97. This is philosopher Chuang Tzu's version of the creation story. It is disputable to accept it as an original myth. However, it is noticeable that the temporal tension is self evident even in this philosophical discourse.

[11] Cf. A. C. Graham, *The Book of Lieh Tzu* (London: John Murray, 1960), pp. 18–19.

[12] Cf. John C. H. Wu, *Lao Tzu* (New York: St. John University Press, 1961), p. 32.

[13] Cf. David Hawkes, *Ch'u Tz'u: The Songs of the South* (Boston: Beacon Press, 1962), pp. 46–47.

[14] Cf. *Shan-hai ching, t'u-tsan, pu-chu* (Taipei: Chung-hua shu-chü, 1960), p. 34.

[15] Cf. Watson, *The Complete Works of Chuang Tzu*, p. 34.

(3) Before Heaven and Earth took shape, there was only undifferentiated formlessness. From this vacuity emerged two gods, who produced the universe (of space and time).[16] The universe produced material force. Everything clear and light drifted up to become Heaven; everything heavy and turbid solidified to form Earth. It was easy for the clear and refined to unite but extremely difficult for the heavy and turbid to solidify. Therefore, Heaven was formed first and the Earth became definite later. The material forces of Heaven and Earth combined to form Yin and Yang. The concentrated forces of Yin and Yang became the four seasons, and the scattered forces of the four seasons became a myriad things. When the hot force of Yang accumulated, fire was produced and the essence of the material force of fire became the sun. When the cold force of Yin accumulated, water was produced and the essence of the material force of water became the moon. The excessive essence of the material forces of fire and water became the stars and planets. Heaven received the sun, the moon and the stars, while Earth received water and soil.[17]

b. The Myth of the Creation of Man

After P'an Ku, Father of all Gods, came a succession of deities: several supernal mother-goddesses and two superlative gods—the Supremacy of Heaven and the Supremacy of Earth.[18] One supernal goddess, Nü Wa, was overcome by the barrenness of the earth and while sitting by a lake she modelled her own images from clay and water.[19] Surprisingly, the clay models came to life with her last touch, and began to sing and dance around her.[20] She was so well pleased that she formed many more images and matched them to engender future generations.[21] Thus man's creation by Nü Wa, the goddess with a dragon

---

[16] Adapted from HNTCSP, cited by Yüan Ke, p. 31. This is the Yin-yang School's version of the creation story. Its authenticity as an original myth is questionable, but a cyclic universe sustained the forces of four seasons is again self evident.

[17] Cf. Chan Wing-tsit, *A Source Book in Chinese Philosophy* (Princeton, New Jersey: Princeton University Press, 1972), pp. 307-08.

[18] Cf. TPYL (Peking: Chung-hua shu-chü, 1960, duplication), I, p. 78-a, b; Jen Ying-ts'ang, *Chung-kuo yüan-ku-shih shu-yao* (Taipei: P'a-mi-er shu-tien, 1954), p. 160.

[19] Cf. Yüan Ke, p. 55.

[20] Adapted from FSTI, cited by TPYL, I, p. 78-4b.

[21] Adapted from FSTI, cited by Ma Su, p. 3. Also cf. Wen I-t'uo, "Kao-t'ang shen-nü ch'uan-shuo chih fen-hsi" in *Shen-hua yü shih* (Peking; Chung-hua shu-chü, 1956), pp. 81-117. Also adapted from LCYL, cited by Yüan Ke, pp. 55-56, 59.

body,[22] heralded the arrival of the third superlative god, the Supremacy of Man. Nü Wa became the Mother of Mothers, and the Goddess of Music and Marriage.[23]

Nü Wa and Fu Hsi

This sketch of a sculpture from a Han tomb of 147 A.D. illustrates the myth of the relationship between the Supernal Mother Goddess and the Sovereign God of the East. The brother-sister or the husband-wife relationship of the two supernal beings is indicated by their coupling bodies. (Cf. *Index of Bronzes and Sculptures of Ancient China*.) Some Chinese legends profess that Nü Wa was a sister of Fu Hsi, the Sovereign God of the East. They were the only two living beings left after the Great Deluge, so they got married and became the primogenitor of future generations.

---

22 *SHC, TT, PC*, p. 189.

23 Adapted from *CWCT*, cited by Yüan Ke, p. 59; from *HNTSLP*, cited by Yüan Ke, pp. 45, 59. Also Cf. Yüan Ke, p. 40.

c. The Myth of the Great Deluge: The Destruction of the Universe

Sometime after the Creation of Man, intolerance between Kung Kung (a mighty God of Water)[24] and his father, Chu Jung[25] (the God of Fire),[26] prompted the former to rise against Chu Jung in a great battle.[27] With the help of Hsiang Liu, a lesser god with nine heads and a serpent's tail,[28] Kung Kung (who himself had a serpent-body and red hair)[29] raised terrible storms and violent floods against his father. The battle was fought from Heaven to Earth and disaster was universal. Kung Kung was soon defeated; in anger and despair he knocked his head on Mount Pu Chou, the Pillar of the Universe, with such force that the four extremes of the Earth collapsed. Accordingly, the dome of the sky was broken and the various fires of Heaven rained upon the Earth; great chasms appeared and floods poured forth, devastating most of the life on Earth.[30]

d. The Myth of Paradise Rebuilt and the Golden Age

The Supernal Mother Goddess, Nü Wa, assumed the task of rebuilding the universe and the earthly paradise.[31] Out of her deep love for mankind, she melted colourful stones to glue the broken sky, used the legs of a giant turtle to support the four posts of the Earth's extremes, and stored reed ashes to fill the chasms and stop the terrible floods.[32] As the physical universe was mended, she killed the devouring beasts, enabling the people of Earth to enjoy a peaceful life again. Eventually life became joyful and happy, and men were free from need and worry. Food was plentiful; all creatures lived jubilantly together—even

---

[24] Adapted from *SCSCPSHPC*, cited by Yüan Ke, pp. 56, 59. Also cf. John Er-wei Tu, *The Mythological System of the Mountain-Sea Classic* (Taipei: Hua-ming shu-chü, 1960), p. 32.

[25] Cf. *SHC, TT, PC*, pp. 137, 211. Also cf. John Er-wei Tu, p. 32.

[26] *SHC, TT, PC*, p. 137.

[27] Adapted from *SCSCPSHPC*, cited by Yüan Ke, pp. 56, 59; from *HNTTWP*, cited by Yüan Ke, p. 67; from *HNTLMP*, cited by Yüan Ke, p. 60. Also cf. *SHC, TT, PC*, p. 189; Shen Yen-ping, pp. 9-10; A. G. Graham, p. 96.

[28] *SHC, TT, PC*, p. 144; also cf. Yüan Ke, p. 56.

[29] *SHC, TT, PC*, p. 189; also cf. Yüan Ke, p. 56; John Er-wei Tu, p. 32.

[30] *SHC, TT, PC*, pp. 36, 189. Cf. Yüan Ke, p. 58; Shen Yen-ping, pp. 9-10. Also adapted from *SCSCPSHPC*, cited by Yüan Ke, p. 59; from *HNTLMP*, cited by Yüan Ke, p. 60; from *HNTTWP*, cited by Yüan Ke, p. 67.

[31] Cf. Yüan Ke, p. 58.

[32] Adapted from *HNTLMP*, cited by Yüan Ke, p. 60; from *SCSCPSHPC*, cited by Shen Yen-ping, p. 10. Cf. Yüan Ke, p. 58; A. G. Graham, p. 96; Shen Yen-ping, p. 9.

the tiger's head and serpent's tail were playful. In fact, men realized no distinction between themselves and nature.[33] It was the Golden Age.[34]

e. The Myth of the Silver Age

Long after the passing of Nü Wa, there came Huang Ti, an almighty god loved and respected by all divine beings. Huang Ti was the first god to be enthroned as the Supremacy of the Universe.[35] At the beginning of his rule and before Titan Ch'ih Yu's cosmic rebellion, the universe was in a fine state of flux and reflux. The day and night and the four seasons entered permanent rotation.[36] There was no conscious differentiation between god, man, and nature. Divine beings were welcome passengers of Earth, the winged gods and goddesses enjoying resorts provided by five giant turtles who bore the five divine mountains on the Ocean-valley of Void.[37] But human beings were the happiest creatures.[38] For example, in the Country of Splendid Flowers (the Hua Hsü Country), in the centre of the Earth, people living in this paradise felt no infection in life nor sorrow for death during their lifespan of some hundred years.[39] The titans, offspring of the gods, flourished immortally and travelled freely between Heaven and Earth through various divine trees or spiral roads on divine mountains.[40] This was the Silver Age.

f. The Myth of Passageways to Heaven

Chinese mythology mentions many passageways to Heaven; the most famous are the spiral roads of Mount K'un Lun,[41] Mount Chao[42] and Mount Teng Pao,[43]

---

[33] Adapted from *HNTLMP* and *HNTPCP*, cited by Yüan Ke, p. 68. Also cf. Yüan Ke, pp. 65-67.

[34] Cf. Yüan Ke, p. 65.

[35] Adapted from *SCLH* and *PT*, cited by Yüan Ke, p. 104. Also cf. Yüan Ke, p. 98.

[36] Adapted from *PHTWH*, cited by Yüan Ke, p. 74; from *HNTLMP*, cited by Yüan Ke, p. 68. Also cf. Yüan Ke, pp. 65, 70.

[37] Adapted from *LTTW* cited by Yüan Ke, p. 74. Also cf. Graham, *The Book of Lieh Tzu*, p. 97; Yüan Ke, pp. 61-63, 67-68, 77, 81.

[38] *SHC, TT, PC*, pp. 135, 141, 177, 185, 192, 206, 80(*TT*).

[39] *SHC, TT, PC*, pp. 76-77, 81, 135, 141, 177, 185, 192, 206, 80(*TT*). Also adapted from *LTHTP*, cited by Yüan Ke, p. 52. Cf. Graham, *The Book of Lieh Tzu*, p. 34.

[40] *SHC, TT, PC*, pp. 206-07; 71, 74, 80(*TT*).

[41] *SHC, TT, PC*, pp. 40-41, 158. Also cf. Yüan Ke, pp. 48, 52.

[42] *SHC, TT, PC*, pp. 160, 206. Also cf. Yüan Ke, pp. 48, 52.

[43] *SHC, TT, PC*, p. 140. Also cf. Yüan Ke, pp. 48, 52.

and the stairway of the divine Chien Tree.[44] Lying in some mysterious region far above all mountains, Mount K'un Lun is held to be the second highest mountain in China. Although not as high as Mount Pu Chou, the Pillar of Heaven,[45] legend describes it as lost in the clouds. It is an abode of perfect blessedness. It consists of five circular regions connected by a spiral road.[46] The first circle, the Mount of Fire, burns forever, making human passage impossible.[47] An Abyss of Sinking Water upon which nothing will float forms the second circle,[48] while the third is the divine palace K'un Lun, formed from five solemn castles and twelve towers.[49] Here the immortals feast on ambrosia and hold conferences.[50] It is entered through a great eastern gate facing the dawn, and guarded by the Daybreak (K'ai Ming), a divine being with a tiger's body and nine heads with human faces.[51] In the fourth circle is the Mount of the Cool Wind. This is the realm of immortalization; the River of Clearness flows into the Lake of Purity, which provides the water of life for immortals' thirst. This sacred water is kept by a lesser god in the shape of a bull with the tail of a horse, eight legs, and two human faces. But he is also the God of War—when he appears there will be war and disaster.[52] The uppermost circle is the Hanging Garden, so high above the clouds that it appears to hang from Heaven. It is the realm of spiritualization, and by ascending this circle one may command the wind and rain.[53] The fifth circle is distinguished as the home of magical trees of Sapphire, Nephrite, Jasper, Jade, Amethyst, Coral, Agate, Amber, Garnet, and Pearl. They provide ambrosia and nectar for the divine beings.[54] The gate to the last circle is guarded by Yin Chao, a mighty god with

---

[44] SHC, TT, PC, pp. 150, 206-07; 71, 74, 79, 80(TT). Cf. Yüan Ke, pp. 48-49, 53.

[45] SHC, TT, PC, pp. 36, 189. Adapted from SCSCPSHPC, cited by Yüan Ke, p. 59; from HNTTWP, cited by Yüan Ke, p. 67. Also cf. Yüan Ke, p. 58.

[46] SHC, TT, PC, pp. 40-41, 158-160. Also adapted from HNTTHP, cited by Yüan Ke, p. 52. Also cf. Yüan Ke, p. 101.

[47] SHC, TT, PC, pp. 193-94. Also adapted from Sou shen chi, XIII, cited by Yüan Ke, p. 105. Also cf. Yüan Ke, p. 101.

[48] SHC, TT, PC, pp. 159, 193-94; 77(TT). Also adapted from Sou shen chi, XIII, cited by Yüan Ke, p. 105. Also cf. Yüan Ke, p. 101.

[49] Adapted from HNTTHP, cited by Yüan Ke, pp. 104, 105. Also cf. Yüan Ke, p. 99.

[50] SHC, TT, PC, pp. 37, 40, 103.

[51] SHC, TT, PC, pp. 37, 40, 103, 159-60. Also cf. Yüan Ke, p. 101.

[52] SHC, TT, PC, p. 39. Also adapted from HNTTHP, cited by Yüan Ke, p. 52. Also cf. Yüan Ke, p. 99.

[53] SHC, TT, PC, p. 39. Also adapted from HNTTHP, cited by Yüan Ke, p. 52. Also cf. Yüan Ke, p. 98.

[54] SHC, TT, PC, pp. 39, 160; 71, 76(TT). Also adapted from HNTTHP, cited by Yüan Ke, p. 104. Also cf. Yüan Ke, p. 99.

a tiger's body, human face, and two tremendous wings.[55] Within the garden several precious trees of life are guarded by Li Chu, a powerful god who has three heads and six eyes. Each head sleeps and watches in turn.[56] Beyond the fifth circle is the divine dwelling of Huang Ti, the realm of divinity and the division between the Nine Heavens and the Nine Earths. Here the gate is kept by Lu Wu, who also has a tiger's body and a human head, as well as nine tails. Lu Wu is also the Superintendent of the Nine Heavens.[57] As a whole the spiral road of Mount K'un Lun is the most solemn and difficult passageway between Heaven and Earth. It is mainly for divine beings and is forbidden to human beings who would surely succumb to its obstacles.[58] With the exception of Mount K'un Lun, the other passageways are at least humanly accessible, however difficult. The spiral roads of Mount Chao and Mount Teng Pao are of this kind.[59]

It is interesting to note that in many of the mythical paradises there is always a divine tree, which bridges Heaven and Earth with numerous branches for the various seekers.[60] The legend of the divine Chien Tree is a perfect example. This tree grows in the centre of Tu Kuang Plain, the central plain of earthly paradise. The land is always green, eternally blooming, and full of singing birds and dancing phoenixes; a land where hundreds of grains ripen in all seasons. The tree itself is very large; its nine spiral branches and nine crooked roots form a stairway to Heaven.[61] It is said that the Sovereign God of the East was the first one to walk through it.[62]

g. The Myth of the Divine Conference and the Rebellion of Titan Ch'ih Yu

During the Silver Age, when Huang Ti was at the height of his power, the first divine conference of gods and spirits was called.[63] It was held atop Mount T'ai (Mount of Great Peace) and the Supremacy of the Universe arrived in a sacred

---

[55] SHC, TT, PC, p. 39. Also cf. Yüan Ke, p. 99.
[56] SHC, TT, PC, p. 160. Also cf. Yüan Ke, p. 100.
[57] SHC, TT, PC, pp. 40–43. Also adapted from HNTTHP, cited by Yüan Ke, p. 52. Also cf. Yüan Ke, p. 98.
[58] SHC, TT, PC, pp. 158; 71(TT).
[59] SHC, TT, PC, pp. 140, 206.
[60] SHC, TT, PC, pp. 140–50, 177, 185, 188, 206.
[61] SHC, TT, PC, pp. 206–07; 74, 80(TT).
[62] SHC, TT, PC, p. 207.
[63] Adapted from HFTSK, cited by Yüan Ke, p. 111. Also cf. Yüan Ke, p. 109.

elephant chariot driven by the holy bird, Pi Fang[64]—a red-striped green ibis with a white beak on a human face and a single leg.[65] His attendant was Titan Ch'ih Yu,[66] who was descended from Yen Ti,[67] the Sovereign God of the South and the God of the Virtue of Fire.[68] Ch'ih Yu possessed a human body, a buffalo horn, four eyes, eight hands, and eight cloven-hoofed legs.[69] He commanded a herd of wolves and tigers to clear the road for the Supreme Being.[70] Dusting the road after Ch'ih Yu were the Elder of the Rain, who boasted the body of a silkworm,[71] and the Chief of the Wind, who had a deer's body with leopard spots and a snake's tail, but a sparrow's head with twin deer horns.[72] Huang Ti examined and then punished or rewarded all the gods and spirits before receiving their praise and respects.[73] Unfortunately, the glory of Huang Ti provoked a strong desire for power in Ch'ih Yu's heart.[74]

Following the Divine Conference, Ch'ih Yu and his eighty-one brothers became ambitious for power.[75] Ch'ih Yu's craving finally resulted in a conspiracy against his grandfather, Yen Ti, the almighty Sun-god in the South.[76] Under Ch'ih Yu's malicious persuasion his titan brothers, the Chief of the Wind and the Elder of the Rain, numerous monsters and evil spirits,[77] the brave people of the south,[78] and the warriors of Miao (offspring of the Supreme Being)[79] all joined the rebellion. In order to avoid a disastrous war

---

[64] Cf. Yüan Ke, p. 111.

[65] *SHC, TT, PC*, pp. 43, 134.

[66] Adapted from *HFTSK*, cited by Yüan Ke, p. 111.

[67] Adapted from *IS*, V, cited by Yüan Ke, p. 74; from *LSHCSCYC*, cited by Yüan Ke, p. 118. Also cf. Yüan Ke, p. 112.

[68] *SHC, TT, PC*, pp. 210-11. Also adapted from *PHTWH* and *HNTSTP*, cited by Yüan Ke, p. 74.

[69] Adapted from *LYHT*, quoted by *TPYL* and from *KT*, quoted by *IS*, cited by Yüan Ke, p. 118. Also cf. Shen Yen-ping, pp. 66-67; Yüan Ke, pp. 112, 118.

[70] Adapted from *HFTSK*, cited by Yüan Ke, p. 111. Also cf. Yüan Ke, p. 109.

[71] Adapted from *HFTSK* and *CTTW*, cited by Yüan Ke, p. 111.

[72] Adapted from *HFTSK* and *CTLS*, cited by Yüan Ke, p. 111.

[73] *SHC, TT, PC*, pp. 38, 57. Also cf. Yüan Ke, pp. 106-07.

[74] Adapted from *CSCMP*, cited by Yüan Ke, p. 118. Also cf. Yüan Ke, p. 112.

[75] Adapted from *LYHT*, quoted by *TPYL*, cited by Yüan Ke, p. 118. Also cf. Shen Yen-ping, pp. 66-67; Yüan Ke, p. 112.

[76] Adapted from *CSCMP*, cited by Yüan Ke, p. 118. Also cf. Yüan Ke, p. 113.

[77] *SHC, TT, PC*, p. 201. Also cf. Shen Yen-ping, pp. 66-67; Yüan Ke, pp. 113, 116.

[78] *SHC, TT, PC*, p. 49. Also adapted from *TYTT*, cited by Yüan Ke, p. 119. Also cf. Yüan Ke, p. 115.

[79] *SHC, TT, PC*, p. 202. Also adapted from *SCLH*, cited by Yüan Ke, p. 89. Also cf. Yüan Ke, pp. 84, 109.

with his beloved human beings, Yen Ti withdrew northwards.[80] Thus Ch'ih Yu occupied the south but, his desire for power unfulfilled, he persuaded Titan K'ua Fu and his followers (offspring of the God of the Soil)[81] to join him in his further rebellion against the Supremacy of the Universe.[82]

The Supreme Being called out all the mysteriously powerful spirits and sacred animals, and the gods and goddesses in all realms to suppress the growing rebellion.[83] The battle was ferocious and it was not until the Goddess of Dark Mystery in the Ninth Heaven devised a mysterious strategy that the Supreme Being was assured of victory and the capture of Ch'ih Yu.[84] All the evil spirits and malicious gods were either killed or deprived of divinity, and their offspring were forever banished from Heaven.[85] Ch'ih Yu died in a set of sacred manacles, which were afterward cast upon a vast plain where they turned into a maple forest. The red leaves of the maple trees forever signify Ch'ih Yu's anger and plead his unfulfilled dream.[86]

h. The Myth of the Estrangement of Earth from Heaven

After the suppression of Ch'ih Yu's rebellion, the Supremacy of the Universe reluctantly ordered his great grandson, Chuan Hsü,[87] to supervise the estrangement of Earth from Heaven to prevent further conspiracies of the mortal against the divine.[88] Upon this order Chuan Hsü despatched two mighty gods: Ch'ung, the God of Life, was sent to close the spiral passageways between Earth and Heaven, and thence to superintend the boundary of the First Heaven; Li was sent to block the stairways to Heaven by divine trees in various paradises, and thence to superintend the great ground and the multitude.[89] Thus

---

[80] Adapted from *CSCMP*, cited by Yüan Ke, p. 118. Also cf. Yüan Ke, p. 113.

[81] *SHC, TT, PC*, pp. 145, 199, 210–11. Also cf. Yüan Ke, pp. 119–20.

[82] *SHC, TT, PC*, pp. 145, 182, 201. Also cf. Yüan Ke, pp. 118, 119, 121.

[83] *SHC, TT, PC*, pp. 182, 201. Also adapted from *CSCMP*, *LYHT* quoted by *TPYL*, and *SCWTPC*, cited by Yüan Ke, pp. 118–19. Also cf. Yüan Ke, pp. 116–17; Shen Yen-ping, pp. 66–67; Ting Ying, *Chung-kuo shang-ku shen-hua ku-shih* (Hong Kong: Shang-hai shu-chü, 1960), pp. 48–56.

[84] Adapted from *HTYNCF*, quoted by *TPYL*, cited by Yüan Ke, p. 125. Also cf. Yüan Ke, pp. 119, 122.

[85] *SHC, TT, PC*, pp. 182, 201. Also adapted from *SCWTPC, SCLH, SCTYM, TWSCCC*, and *MTFK*, cited by Yüan Ke, pp. 125–26. Also cf. Yüan Ke, p. 122; Shen Yen-ping, pp. 66–67.

[86] *SHC, TT, PC*, p. 186. Also cf. Yüan Ke, pp. 122–23, 126.

[87] *SHC, TT, PC*, pp. 192, 205–06. Also cf. Yüan Ke, p. 83.

[88] Adapted from *SCLH*, cited by Yüan Ke, p. 89.

[89] *SHC, TT, PC*, p. 192. Also adapted from *KYCY*, cited by Yüan Ke, p. 89. Also cf. Yüan Ke, pp. 84–85.

the great rebellion resulted in the separation of the mortal from the divine, and the further remoteness between Heaven and Earth.[90]

i. The Myth of Disasters on Earth After the Estrangement

After the estrangement from Heaven, the Earth was damned with fallen gods, titans, and warriors, as well as numerous cursed animals. They occupied deserted mountains, forests, and marshes to haunt nearby tribes or villages. Mankind suffered not only from mortality, but also from the miseries inflicted by these fallen creatures.[91] To name a few: droughts were the curse of the six-legged, four-winged Fei I Snake;[92] floods, the legacy of the Ling Ling Beast, which had a bull-body with tiger stripes;[93] the Fei Beast with his ox-body, white head, single eye and snake tail caused plagues;[94] while the snake-bodied, four-winged, three-legged, and six-eyed Suan Yu Bird brought famines.[95] Tempests were the curse of Chi Meng, which had a dragon's head upon a human body;[96] and finally the fallen god Keng Fu brought destruction wherever he visited.[97]

j. The Myth of the God of Time and the Journey of the Sun

Following the Earth's estrangement from Heaven, and with the appointment to be the God of the Earth, Li descended to Earth and begot I (or I Ming), a son whom he assigned to be the God of Time. I Ming dwells on the top of the Sun-moon Mountain in the extreme western region and guards the sacred Wu Chi Door to Heaven; this is the final earthly station of the journeys of the sun and moon. His other tasks include regulating the order and courses of all stellar phenomena.[98]

Ti Chün, a great-grandson of the Supreme Being and a brother of Chuan Hsü,[99] was once the Sovereign God of the East.[100] He has a horned bird-head

---

[90] Adapted from SCLH and KYCY, cited by Yüan Ke, p. 89.
[91] Adapted from SCLH, cited by Yüan Ke, p. 89. Also cf. Yüan Ke, pp. 91–94.
[92] SHC, TT, PC, p. 25.
[93] SHC, TT, PC, p. 78.
[94] SHC, TT, PC, p. 84.
[95] SHC, TT, PC, p. 66.
[96] SHC, TT, PC, p. 109.
[97] SHC, TT, PC, p. 119.
[98] SHC, TT, PC, p. 192. Also cf. Yüan Ke, p. 85.
[99] SHC, TT, PC, pp. 177, 184, 192, 205–06. Also cf. Yüan Ke, p. 82, 145, 151.
[100] Cf. Yüan Ke, p. 141.

and the body of an orangutan, but has only one leg and thus leans on a staff much of the time. Ti Chün lives in the desolate east and has become friendly with the colourful phoenixes of the region.[101] His two wives are of great importance: one, Ch'ang Hsi, gave birth to twelve moons and was named the Goddess of the Moon; the other, Hsi He, bore ten suns and became the Goddess of the Sun.[102]

Hsi He lives with her family in Valley T'ang, which is an ocean valley of boiling water and is located beyond the Great Sea of the East.[103] Her sons, the ten suns, in the shape of a three-legged bird whose golden feathers radiate heat and light, rest on the divine giant Fu Sang Tree and each is radiantly active in turn.[104] It is said that whenever one sun returns from the firmamental journey, the sacred Jade Cock atop the Fu Sang Tree crows joyously.[105] The next sun on duty will immediately bathe in Valley T'ang, causing the ocean to surge and boil into the flood tide. And the Golden Pheasant on top of the Sacred Peach Tree of the Hell will crow to close the gates of the Dark Kingdom, and all the earthly roosters will crow to announce the arrival of a new sun and a new day.[106]

The Goddess of the Sun rides a brilliant chariot pulled by six winged dragons, while the sun emits thousands of bright rays in every direction.[107] As the chariot passes through the sky, it crosses several stations which indicate different periods of the day. It does not stop until reaching the Spring of Sorrow, where Hsi He leaves her son for the remainder of the journey and drives the chariot back to Valley T'ang to rest before the next voyage. Each son's departure evokes such grief in Hsi He that her tears form the Spring of Sorrows. From the point of leave-taking the sun walks alone to the Vast Abyss Yü, where he bathes to wash off the dust of the journey. This action again causes the tides to rise. Then the sun descends slowly into Meng, the Valley of Chaos, where he reports to the God of Time, who in turn opens the Wu Chi Door to allow the sun return to Heaven.[108]

---

[101] *SHC, TT, PC*, pp. 178, 181, 184. Also cf. Yüan Ke, pp. 77, 81-82, 141-45, 147.

[102] *SHC, TT, PC*, pp. 188, 193. Also cf. Yüan Ke, pp. 142, 173.

[103] *SHC, TT, PC*, pp. 45, 188. Also cf. Yüan Ke, p. 173.

[104] *SHC, TT, PC*, pp. 150, 181. Also adapted from *CTTW* AND *HNTCSP*, cited by Yüan Ke, p. 185. Also cf. Yüan Ke, p. 180.

[105] Adapted from *SIC*, cited by Yüan Ke, p. 178. Also cf. Yüan Ke, p. 173.

[106] Cf. Yüan Ke, pp. 178, 173.

[107] *SHC, TT, PC*, pp. 181, 188. Also adapted from *CTLS*, cited by Yüan Ke, p. 178. Also cf. Yüan Ke, pp. 173-74.

[108] *SHC, TT, PC*, pp. 181, 192. Adapted from *HNTTWP*, cited by Yüan Ke, p. 178. Also cf. Yüan Ke, pp. 120, 174; Lao Kan, "The Division of Time in the Han Dynasty as Seen in the Wooden Slips," *Bulletin of the Institute of History and Philology* (Academia Sinica, Taipei), 49-1, (1969), p. 292.

Ten Suns on the Cosmic Tree

This sketch is taken from "the Flying Garment" found atop the innermost casket of the tomb of Lady Ch'eng, a consort of Emperor Ch'in, who reigned from 156 to 141 B.C. The sketch shows ten suns resting on the cosmic Fu Sang Tree, the divine mulberry tree.

k. The Myth of Titan P'eng Tsu's Sorrow Over the Ephemerality of Life

Titan P'eng Tsu was a descendant of the Sovereign God of the North, and was unique in being born from his mother's armpit.[109] He lived longer than any earth-bound mortal, and at the age of eight hundred years he was asked by King Yin's errand-girl about the meaning of long life and happiness. P'eng Tsu answered: "Life is short, and yet so full of despair, sorrow, anguish, anxiety, worry, departure and death. It is fleeting and I am afraid I shall die soon. I cannot teach you the way of long life. Forgive me, but now I must leave."

---

[109] Adapted from *SPSHP*, cited by Yüan Ke, p. 90. Also cf. Yüan Ke, p. 87.

With a profound sigh, he paced westward into the wilderness.[110] It is said that Titan P'eng Tsu had been playing tricks with Death and had always been travelling westward to seek for immortality.[111] To the extreme west, beyond the Dark River,[112] there lies the Great Crescent of Sinking Sand,[113] marking the boundary between Earth and Heaven.[114] West of it there stand three sacred mountains: Mount K'un Lun with a spiral road to divinity and various kinds of ambrosia;[115] Mount Ch'ang Liu, the abode of the God of Completion who supervises the sun to go down, and Mount Yü, the abode of the God of Perfection who superintends the golden sunset.[116] Many years later the dying titan was seen on the western boundary of the Great Crescent of Sinking Sand, riding an old camel towards the evening horizon in search of the sunset's embrace.[117]

1. The Myth of Titan K'ua Fu, Who Chased the Sun

The K'ua Fu Titans were offspring of Hou T'u, the God of Earth[118] and the Dark Kingdom of Hades.[119] These titans had human bodies with yellow snakes hanging from each ear and hand.[120] After the banishment they lived with their descendants in the extreme northern territory of the Earth, on the range of the northern Pillar of the Universe.[121] Among them lived the brave and well-liked K'ua Fu. He was famous for his great ambition to catch the Sun and thus challenge time to stop. On huge legs he chased the sun to its point of descent by the brink of Abyss Yü. Extending his hands to catch the fleeting chariot of brightness, he was overcome by thirst and bent over the Yellow and Wei Rivers to drink. Both rivers were drained in a swallow, yet his thirst was not quenched. In desperation, K'ua Fu interrupted his chase and headed northwest for the Great Marshes, where there were thousands of square miles of water. Un-

---

[110] Adapted from *CTTW* and *SHC*, cited by Yüan Ke, p. 90.
[111] Adapted from folktales of southern China. Also cf. Yüan Ke, p. 87.
[112] *SHC, TT, PC*, pp. 205-06.
[113] *SHC, TT, PC*, pp. 205-06; 20(*PC*).
[114] *SHC, TT, PC*, p. 75(*TT*).
[115] *SHC, TT, PC*, pp. 158-60.
[116] *SHC, TT, PC*, pp. 43, 45.
[117] Adapted from *SHC*, cited by Yüan Ke, p. 90. Also cf. Yüan Ke, p. 87.
[118] *SHC, TT, PC*, p. 145.
[119] *SHC, TT, PC*, p. 209. Also adapted from *CTCH*, cited by Yüan Ke, pp. 111, 125; from *LTTW*, cited by Yüan Ke, p. 139. Also cf. Yüan Ke, pp. 119-20.
[120] *SHC, TT, PC*, pp. 145, 199.
[121] *SHC, TT, PC*, p. 199.

rtunately, he died of thirst before reaching the water's edge. As his gargantuan body crashed to the ground, the earth trembled with thunder echoing his grief.[122] His staff changed into a vast peach forest.[123]

m. The Myth of Archer God Hou I, Who Shot Down Nine Suns

It is said that, despite the regular route of their celestial journey and the strict order of their rotation, the Ten Suns once rushed out of Valley T'ang without their mother's escort and romped over the firmament in defiance of her warnings and commands.[124] The deliberate order of time was thus disintegrated, the tremendous heat and light of ten suns caused a severe drought, and giant monsters fled the burning forests and the boiling marshes to trample upon the good earth. The human race suffered terribly and the Great King Yao appealed to the Supremacy of Heaven for deliverance.[125] In answer the Supreme Being sent his sacred white bow and red arrows to Hou I, the greatest archer in Heaven,[126] with instructions that he was to superintend the ten suns back to their dutiful regulation of time.[127] But when Hou I saw the great devastation on Earth his sympathies lay with humanity, and in a great rage he killed nine of the suns and slightly wounded the last one before it fled back to its usual post.[128] Then the mighty Archer God slew seven of the largest and most dangerous monsters and presented the gigantic swine in sacrifice to the Supreme Being. Yet the Supremacy of Heaven was not pleased[129] and the Archer God and his wife, Ch'ang O (a fairy goddess) were deprived of their divinity and banished from Heaven. They descended to Earth and later became beloved king and queen among mankind.[130]

---

[122] *SHC, TT, PC*, pp. 145, 199. Also adapted from *CTLS*, cited by Yüan Ke, p. 125. Also cf. Yüan Ke, p. 120.

[123] *SHC, TT, PC*, pp. 101, 145. Also cf. Yüan Ke, p. 125.

[124] Adapted from *CTCH*, cited by Yüan Ke, p. 177. Also cf. Yüan Ke, p. 174.

[125] *SHC, TT, PC*, p. 140. Also adapted from *HNTPCP*, cited by Yüan Ke, p. 177. Also cf. Yüan Ke, pp. 175–78.

[126] Adapted from *CTSKC, HFTSL*, and *HNTHWP*, cited by Yüan Ke, p. 179. Also cf. Yüan Ke, p. 177.

[127] *SHC, TT, PC*, p. 210. Also adapted from *CTTW*, cited by Yüan Ke, p. 179. Also cf. Yüan Ke, p. 177.

[128] Adapted from *HNTCSP, HNTPCP*, and *CTTW*, cited by Yüan Ke, p. 185. Also cf. Yüan Ke, p. 180.

[129] Adapted from *CTTW, HNTPCP*, cited by Yüan Ke, pp. 185–86. Also cf. Yüan Ke, pp. 180–83.

[130] Adapted from *HNTLMP*, cited by Yüan Ke, p. 184; from *CTLS*, cited by Yüan Ke, p. 186. Also cf. Yüan Ke, pp. 179, 184.

24   A Comparative Study of Chinese and Western Cyclic Myths

The Archer God Shooting the Ten Suns

This sketch of a funerary sculpture from a Han tomb of 147 A.D. shows that Archer God Hou I was shooting the solar birds on or around the cosmic Fu Sang Tree. (Cf. E. Chavannes, *Mission Archeologique dan la Chine Septentrionale*, pl. li. Also cf. Olov R. T. Janes, Archaeological Research in Indo-China, V.11, 1951, p.47)

n. The Myth of the Flight to the Moon

Following her banishment to Earth, Ch'ang O (a fairy sister of the Goddess of the Moon)[131] became deeply nostalgic for divinity of Heaven and dreaded the dark kingdom of Death.[132] So great was her grief that King Hou I decided to make a pilgrimage to Mount K'un Lun to plead with Hsi Wang Mu, a Supernal Mother Goddess, for the pill of immortality.[133] With his semi-divine powers and heroic perseverance, King Hou I overcame the great obstacles along the way. He passed the Mount of Fire, the Abyss of Sinking Water, and the guards

---

[131] Cf. Yüan Ke, pp. 179, 184.
[132] Cf. Yüan Ke, p. 195.
[133] Cf. Yüan Ke, p. 195.

of sacred animals and mighty gods in each circle.[134] At the Lake of Purity he paid homage to the Supernal Goddess and pleaded his case. Touched by his great strength of will, Hsi Wang Mu generously gave him two sacred pills; one pill provided immortality, but two promised divinity.[135] Upon Hou I's return, Ch'ang O was overjoyed at the shared prospect of immortality. But her desire for divinity and reconciliation with Heaven was so strong that she betrayed her husband and swallowed both pills. Ch'ang O thereupon regained divinity and was immediately transported to the moon.[136] However, Ch'ang O was so ashamed of her disloyalty that she fled silently to the Palace of the Moon. There she found only a sacred rabbit, which was grinding capsules and ointments in eternal penance for neglect of duty; and Wu Kang, a lesser god, who was hewing the Sacred Laurel Tree which immediately healed itself after each chop. He was performing this endless toil of punishment for misbehaviour in Heaven. It is said that in the Palace of the Moon Ch'ang O lived her divine immortality with some regret.[137]

Hsi Wang Mu dwells in several places: Mount Jade (west of Mount K'un Lun), the Lake of Purity in the fourth circle of K'un Lun, and the Valley Meng (the Valley of Chaos) in the western extreme.[138] As one of the Supernal Mothers of Gods in Heaven, she appears as a graceful goddess. However, as the Goddess of Plague and Punishment,[139] she has extremely long hair, tiger fangs, and a panther's tail.[140] She prefers to live among the animals and is constantly attended by three huge birds-of-prey.[141] In addition she is in charge of the divine trees of life and immortality in the Hanging Gardens.[142] These trees bear fruit every thousand years, and the fruits are guarded by Li Chu. The Supernal Goddess collects the fruit as the source of the ambrosial pills of immortality.[143]

---

[134] *SHC, TT, PC*, pp. 159, 194. Also cf. Yüan Ke, p. 197.

[135] *SHC, TT, PC*, pp. 160; 76(*TT*). Also cf. Yüan Ke, pp. 197. Also adapted from *HNTLMP*, cited by Yüan Ke, p. 200.

[136] Adapted from *HNTLMP*, cited by Yüan Ke, p. 200. Also cf. Eberhard, Wolfram, *FOLKTALES OF CHINA* (Chicago: The University of Chicago Press, 1963), p. 125; Yüan Ke, pp. 197-99, 200.

[137] Adapted from *HNTLMP*, cited by Yüan Ke, p. 200. Also cf. Yüan Ke, pp. 199-200.

[138] *SHC, TT, PC*, p. 42. Also adapted from *MTTC*, cited by Yüan Ke, p. 200. Also cf. Yüan Ke, p. 197.

[139] *SHC, TT, PC*, p. 42. Also cf. Yüan Ke, p. 196.

[140] Cf. Yüan Ke, p. 195.

[141] *SHC, TT, PC*, pp. 163, 44, 191. Also cf. Yüan Ke, p. 196.

[142] Cf. Yüan Ke, pp. 196-97.

[143] Cf. Yüan Ke, p. 196.

The Flight To The Moon

This sketch is taken from "the Flying Garment" found atop the innermost casket of the tomb of Lady Ch'eng, a consort of Emperor Ch'in, who reigned from 156 to 141 B.C. The picture shows that Goddess Ch'ang O was fleeing to the moon on a dragon's wings, after she swallowed the elixir of immortality.

0. The Myth of the Divine Administration

Huang Ti, the omnipotent and omnipresent Supreme Being, the Supremacy of the Universe,[144] has four faces which enable him to see in the four directions all dimensions, all spirits, and all beings.[145] In constant supervision, he dwells in the centre of the Universe and is also the Sovereign God of the Centre, the fifth direction. His sacred animal is a yellow dragon, symbol of the earth. The assistant to the Sovereign God of the Centre is called Hou T'u (the Great Earth) or Chü Lung (the Horned Dragon).[146] He is the son of Kung Kung.[147] He

---

[144] Adapted from *SCLH* and *PT*, cited by Yüan Ke, p. 104.
[145] Cf. Yüan Ke, pp. 106, 110.
[146] Adapted from *HNTTWP*, cited by Yüan Ke, p. 110. Also cf. Shen Yen-ping, p. 102; Ting Ying, pp. 42–47; Yüan Ke, p. 106.

has a bull's body with a three-eyed tiger's head and two dragon horns.[148] While on assistance he holds a cord or rope-rule to regulate the four directions of the universe.[149] He is thus the God of Earth and the God of Soil. He is also the God of Hades, the dark kingdom of spirits and souls.[150]

Under the Supreme Being, there are four sovereign gods helping him supervise the entire universe. The Sovereign God of the East is called T'ai Hao (the Great Spirit of Vitality).[151] He is the high vigour of the rising sun in the morning and the great vitality of the east wind dissipating the winter's severe cold. He is the God of the Virtue of Wood.[152] He has a dragon's body and his sacred animal is a green dragon, symbol of vitality and life.[153] The Sovereign God of the East is also called Fu Hsi (the Great Herder). The classics profess that he is a brother of Supernal Mother Goddess Nü Wa. They were the only two living beings left after the Great Deluge, so they got married and became the primogenitor of future generations.[154] The Assistant to the Sovereign God of the East is called Chü Mang (the Bud and Sprout) or Ch'ung (the Multiple and Regeneracy).[155] He is the son of the Sovereign God of the West and the brother of the God of Autumn. He has a square face and a bird's body.[156] He is the mighty god who once executed the Estrangement of Earth from Heaven and thence superintends the first circle of Heaven.[157] While on assistance, he rides two dragons and holds a pair of compasses to regulate the course of the spring. He is thus the God of the Spring,[158] the God of Wood, and the God of Life.[159]

---

[147] Cf. Morohashi, *Dai kanwa jiten*, pp. 744, 746 (v. 2).

[148] *SHC, TT, PC*, pp. 199, 209. Also adapted from *CTCH*, cited by Yüan Ke, p. 125.

[149] Adapted from *HNTTWP*, cited by Yüan Ke, p. 110. Also cf. Shen Yen-ping, p. 102; Yüan Ke, p. 106.

[150] *SHC, TT, PC*, pp. 133, 209. Also adapted from *CTCH*, cited by Yüan Ke, p. 111. Also cf. Yüan Ke, p. 107.

[151] Adapted from *HNTTWP*, cited by Yüan Ke, p. 53. Also cf. Shen Yen-ping, p. 100.

[152] Adapted from *LSCCMCC*, cited by Shen Yen-ping, p. 102.

[153] Adapted from *HNTTWP*, cited by Shen Yen-ping, p. 102. Cf. Yüan Ke, pp. 45, 52.

[154] See page 13 for the sketch of a sculpture of Nü Wa and Fu Hsi from a Han tomb of 147 A.D. It indicates their relationship with coupling bodies.

[155] Adapted from *HNTTWP*, cited by Yüan Ke, p. 53. Also cf. Shen Yen-ping, p. 100.

[156] Adapted from *LSCCMCC*, cited by Yüan Ke, p. 53.

[157] *SHC, TT, PC*, p. 192. Also adapted from *KYCY*, cited by Yüan Ke, p. 89. Also cf. Yüan Ke, pp. 84–85.

[158] *SHC, TT, PC*, pp. 152; 73(*TT*). Also adapted from *HNTTWP*, cited by Yüan Ke, p. 53. Also cf. Shen Yen-ping, p. 102; Yüan Ke, p. 106.

[159] *SHC, TT, PC*, pp. 73, 152. Also adapted from *LCYL* and *LSCCMCC*, cited by Yüan Ke, p. 53. Also cf. Yüan Ke, pp. 53, 79.

28   *A Comparative Study of Chinese and Western Cyclic Myths*

The Sovereign God of the South is called Yen Ti (the Great Heat and Light). He is the burning heat and the brilliant light brought about by the south wind, which follows grain-rain and encourages growth. He is an almighty Sun-god in the south. He is the God of the Virtue of Fire[160] and the God of Agriculture.[161] He has an ox body[162] and his sacred animal is a red sparrow, symbol of multiplicity.[163] The Assistant to the Sovereign God of the South is called Chu Jung (the Persistent Heat and Light of Fire) or Li (the Multitude). He is the jubilant heat and light of the summer solstice, which drives living things to growth. He is a great great-grandson of Chuan Hsü, the Sovereign God of the North. He is the mighty god who once executed the Estrangement of Earth from Heaven and thence superintends the great ground and the multitude. He has a beast's body. While on assistance, he rides two dragons and holds a yoke or a beam to regulate the course of the summer. He is thus the God of the Great Ground, the God of the Multitude, the God of the Summer, the God of Fire, and the God of Growth.[164]

The Sovereign God of the West is called Shao Hao (the Small Spirit of Vitality). He is the vigour and vitality, which will be dormant in the seed when the west wind brings forth sweet ripeness. He is the God of the Virtue of Gold or Metal.[165] He is the son of the Supreme Being. He has a falcon's body[166] and his sacred animal is a white tiger, symbol of the inexorable beast of prey.[167] He is also known as Pai Ti (the God of the White Star), and is sometimes called Chin T'ien (the Golden Sky). He also dwells on the top of Mount Ch'ang Liu to supervise the sun-down and is called Yüan Shen (the God of Completion and Profundity).[168] The Assistant to the Sovereign God of the West is Called Ju Shou (the Reaper of the Roots of Exuberant Grass) or Kai (Necessity and Destiny).[169] He is the collector of the beard. He is the necessity and destiny for things reaching completion and perfection. He is the second son of the Sovereign God of the West and the brother of the God of Wood, the God

---

[160] Adapted from *HNTTWP*, cited by Yüan Ke, p. 110; from *HNTTWP*, cited by Shen Yen-ping, p. 102; from *HNTSTP*, cited by Yüan Ke, p. 74.
[161] Adapted from *PHTWH* and *HNTSTP*, cited by Yüan Ke, p. 74.
[162] Adapted from *TWSC*, quoted by *IS*, cited by Yüan Ke, p. 74.
[163] Adapted from *HNTTWP*, cited by Shen Yen-ping, p. 102.
[164] *SHC, TT, PC*, pp. 137, 210–11; 66(*TT*). Also adapted from *HNTTWP*, cited by Shen Yen-ping, p. 102. Also cf. Yüan Ke, p. 106.
[165] Adapted from *HNTTWP*, cited by Shen Yen-ping, p. 102; from *HNTSTP*, cited by Yüan Ke, p. 82.
[166] Cf. Yüan Ke, pp. 77, 82.
[167] Adapted from *HNTTWP*, cited by Shen Yen-ping, p. 102.
[168] *SHC, TT, PC*, p. 43. Also cf. Yüan Ke, pp. 76, 82.
[169] *SHC, TT, PC*, pp. 66, 142.

of Life and the God of the Spring.[170] He has a tiger's body with white hair, and snakes hanging from each ear. While on assistance, he rides two dragons and holds a square to regulate the course of the autumn. He also dwells on the top of Mount Yü to superintend the sunset and is called Hung Kuang (the Red Light).[171] He is the God of Gold or Metal, the God of the Autumn,[172] the God of Harvest and the God of Punishment.[173]

The Sovereign God of the North is called Chuan Hsü (the Cautious Refraining). He is the abstention of the earth after the cold dew. He is the God of the Virtue of Water.[174] He is a grandson of the Supreme Being and is the mighty god who once executed the Estrangement of Earth from Heaven. He has a unicorn's body[175] and his sacred animal is a black tortoise, symbol of the mystery of hibernation.[176] The Assistant to the Sovereign God of the North is called Yü Ch'iang (the Latent Vitality of Darkness and Chaos).[177] He is sometime called Yüan Meng (the Origin in the Occult Darkness) or Hsüan Meng (the Latent Profundity and Perseverance in the Occult Chaos).[178] He is the North-west Wind, given birth by the spirit of the Pu Chou Wind from the northern Pillar of the Universe.[179] As the God of the North-west Wind, he has a bird's body with green snakes hanging down from each ear, and a pair of tremendous wings which can raise a tempest of plague of death in a single stroke. And As the God of the North Sea, he has a black face and a fish body capable of transforming into a gigantic phoenix, which brings along water from the north for the spring. While on assistance, he rides two dragons and holds a weight to regulate the course of the winter.[180] He is thus the God of Water, the God of the Winter and the God of Hibernation.[181]

---

170 Adapted from *KYCY*, cited by Yüan Ke, p. 82. Also cf. Yüan Ke, p 78.

171 *SHC, TT, PC*, pp. 45, 142; 66(*TT*). Also cf. Yüan Ke, p. 106.

172 Adapted from *HNTTWP*, cited by Yüan Ke, p. 110.

173 *SHC, TT, PC*, p. 66. Also adapted from *KYCY*, cited by Yüan Ke, p. 82.

174 Adapted from *HNTTWP*, cited by Shen Yen-ping, p. 102; from *HNTSTP*, cited by Yüan Ke, p. 89.

175 *SHC, TT, PC*, pp. 192, 205–06. Also adapted from *SCLH* and *KYCY*, cited by Yüan Ke, pp. 83, 89.

176 Adapted from *HNTTWP*, cited by Shen Yen-ping, p. 102.

177 *SHC, TT, PC*, p. 147. Also adapted from *HNTTWP*, cited by Shen Yen-ping, p. 102.

178 *SHC, TT, PC*, p. 147. Also adapted from *HNTSTP*, cited by Yüan Ke, p. 89; from *HNTTWP*, cited by Shen Yen-ping, p. 102.

179 Adapted from *HNTTHP* and *LSCCYSP*, cited by Yüan Ke, p. 67.

180 *SHC, TT, PC*, pp. 72, 147, 180; 72(*TT*). Also cf. Yüan Ke, pp. 62–63, 67.

181 Adapted from *HNTTWP*, cited by Shen Yen-ping, p. 102; Also cf. Yüan Ke, p. 106.

Direction	Colour	Season	Element	Cyclic Nature	God of Direction	Assistant God
Centre	Yellow	Mid-Season	Soil	The Centre	Huang Ti	Hou T'u or (Chu Lung)
East	Green	Spring	Wood	Birth or (Rebirth)	T'ai Hao	Chü Mang or (Ch'ung)
South	Red	Summer	Fire	Growth	Yen Ti	Chu Jung
West	White	Autumn	Metal or (Gold)	Harvest	Shao Hao	Ju Shou or (Kai)
North	Black	Winter	Water	Death or (Hibernation)	Chuan Hsü	Yü Ch'iang

God	Sacred Animal	Denotation And Connotation	Description And Attribution
Huang Ti	-The Yellow Dragon -Symbol of the earth	-The Supremacy of the Universe -The Omnipotent Supreme Being	-Four faces to oversee all directions -The Sovereign God of the Center
T'ai Hao	-The Green Dragon -Symbol of great vitality	-The great spirit of the rising sun of the morning and the east wind	-A dragon body -The God of the Virtue of Wood
Yen Ti	-The Red Sparrow -Symbol of multiplicity	-The burning heat and the brilliant light brought about by the south wind	-An ox body -The God of the Virtue of Fire
Shao Hao	-The White Tiger -Symbol of beast of prey	-The low or small spirit of vigor and vitality dormant in the seed	-A falcon body -The God of the Virtue of Metal or Gold
Chuan Hsü	-The Black Tortoise -Symbol of hibernation	-The cautious refraining -The abstention of earth after the cold dew	-A unicorn body -The God of the Virtue of Water

Aspects of Cyclic Myth in Chinese Mythology   31

Assistant	Attribution	Denotation and Connotation	Description
Hou T'u (Chü Lung)	-The God of Earth -The God of Hades	-The Great Earth -The horned dragon	-A bull's body with a three-eyed tiger head and two dragon horns -Riding on two dragons, holding a cord or rope-rule to regulate the four directions of the universe
Chü Mang (Ch'ung)	-The God of Wood -The God of the Spring -The God of Life	-The bud and sprout in the spring -The multiplication and regeneracy	-A square face and a bird body -Riding on two dragons, holding a pair of compasses to regulate the course of the spring
Chu Jung (Li)	-The God of Fire -The God of the Summer -The God of Growth -The God of the Multitude	-The persistent heat and light of fire -The jubilant heat and light of the summer solstice	-A beast body -Riding on two dragons, holding a yoke or a beam to regulate the course of the summer
Ju Shou (Kai)	-The God of Metal -The God of the Autumn -The God of Harvest -The God of Punishment	-The reaper of exuberant grass -The collector of the beard -The midwife for accouchment -Completion and destination	-A tiger body with white hair and with snakes hanging down from each ear -Riding on two dragons, holding a square to regulate the course of the autumn
Yü Ch'iang (Yüan Meng)	-The God of Water -The God of Wind -The God of North Sea -The God of the Winter	-The latent vitality in darkness -The origin in the occult chaos -At ease with power and vigor -The latent perseverance	-A bird body with a green snake hanging down from each ear, a pair of huge wings; or a fish body -Riding on two dragons, holding a weight to regulate the course of the winter

2. The Temporal Envisagement in Chinese Mythology

Perhaps more so in China than in any other culture of comparable antiquity, myth circumscribes a universe in which man's psychological, philosophical, religious, social, and even political histories are centripetally anchored in abstractions of cyclic renewal. Using the time-honoured principle of divide and conquer we trace three great roots of ancient Chinese culture—mythology, ritual, and philosophy—to discern their culmination into a cyclic myth which remains one of the foundations of Chinese life.

To discover the origins of the Chinese cyclic myth is to reveal the existential orientation of most primitive peoples. To restate briefly the general formation of the cyclic myth, in the world there is everywhere tangible evidence of recurrent rhythm—the periodic alternation of tides, the solar and lunar cycles, and seasonal variation. Against this cosmic rhythm is placed the undeniable linearity of human consciousness. To alleviate the intolerable tensions of this disparity, as Northrop Frye perceives it, myth seizes on the fundamental element in nature and assimilates it to the human cycle.[1] From the periodicity of nature primitive man derives a cyclical notion of time subsumed in a cyclic mythology. It is through this premise that we trace the temporal sense as an essential experience of human existence through Chinese myth to its eventual dramatization in the Myth of Divine Administration.[2]

Only by emphasizing the process through which man reveals the tensions which compelled the formation of the original myth can we adequately explain the time-sense of early Chinese mythology. What later became the developed cosmogonic cyclic myth underlying Chinese culture was originally fragmented into separate myths of creation, rebellion, and reconciliation. The cyclic myth may be seen as the developed response to the tensions which these earlier myths portrayed.

a. The Myth of the Estrangement of Earth from Heaven: A Myth of the Rise of Time and Mortality

Chinese mythology in one aspect is the continuous story of maturing temporal awareness; the harmony of time and its frustrations, the ripeness of time, and the urge toward rebellion against or reconciliation with time. In the Chinese myth of creation the cosmic cycles are but universal courses necessary for great harmony. Light and darkness, seasons, and tides were no cause for concern for

---

[1] Cf. Frye, *Fables of Identity*, p. 32.

[2] The text of the Myth of Divine Administration and other myths mentioned in this chapter are provided in the "Reconstruction of Fifteen Chinese Myth".

the earliest men in myth for they were still in some sense divine; although separated by P'an Ku, Heaven and Earth remained in communication, and divine and mortal beings mingled freely. The divine and human worlds further coincided through cosmic mountains and cosmic trees which joined Heaven and Earth. There was no conscious differentiation between god, man, and nature. Mankind enjoyed the paradisal rewards of a thousand-year lifespan which could be extended by partaking of sacred fruits and waters along the pathways to Heaven. Existence, in a word, was eternal and it was not until the estrangement of Earth from Heaven that temporal tensions arose.

In "A Classification of Shang and Chou Myths", Chang Kuang-chih suggests that the accessibility of the world of god to the world of man was taken for granted throughout the Shang (1766–1123 B.C.) and the Chou (1122–249 B.C.) periods; but in later periods in some traditional versions the communication between the two worlds was completely severed.[3] In fact, the Myth of Estrangement of Earth from Heaven had early appeared in *The Book of Documents*, supposedly compiled in the Spring and Autumn period (722–481 B.C.) and was at once available in several important classics of the period such as *The Remarks Concerning the States*, *The Bamboo Books*, and *The Mountain and Sea Classic*.[4]

The temporal harmony enjoyed by the man-gods during the Golden Age was soon shattered by the Myth of the Rebellion of Ch'ih Yu. Immediately evident are parallels between the causes and effects of Ch'ih Yu's rebellion and the Christian myths of Satan's fall and Eve's disobedience. Although both descents are based on the sin of pride, the separation of the worldly from the divine is more pronounced in the Christian view, and less humanly-directed in the Chinese version. Thus Titan Ch'ih Yu is descended from Yan Ti, one of the Sun-Gods, and possesses the fantastic physical attributes more typical of the divine than the mortal. We must agree with T'ang Chün-i that one of the characteristics of Chinese culture is the absence of an abyssal differentiation

---

[3] Cf. Chang Kuang-chih, "A Classification of Shang and Chou Myths," *Bulletin of the Institute of Ethnology, Academia Sinica*, 14 (Autumn 1962, Taipei), p. 83.

[4] *The Book of Documents* which chiefly consists of addresses to and from the throne, supposedly beginning with the period of the legendary Emperors Yao and Shun (circa 3000 B.C.), was first compiled in the Spring and Autumn period (722–481 B.C.) and then was destroyed during the "Burning of the Books" in 220 B.C. It was later restored and edited in the Western Han dynasty by K'ung An-kuo (156–74 B.C.). *The Remarks Concerning the State of Ch'u* was compiled circa 400 B.C. The present version of *The Mountain and Sea Classic* was not compiled until after the beginning of the Christian era, but it contains myths and legends which had their origins in at least the Chou dynasty (1122–249 B.C.). Scholars believe that its original version was compiled in the Warring States period, circa 372 B.C. *The Bamboo Books* supply a condensed record of reigns and events supposedly from 2700 B.C. to 300 B.C. and was discovered in 279 A.D. in the grave of Duke Hsiang of Wei, who died in 294 B.C.

between human and divine beings,[5] and this in time furthers the emerging concept of the continuing (though diminished) unity of Heaven and man. The Chinese myths support Mircea Eliade's theory that primitive peoples regard the existence of man in the cosmos as a fall from a spontaneously continuous present.[6] To this end Ch'ih Yu's rebellion resulted in the experience of time as the support and limitation of existence; it was a deprivation of immortality and a denial of eternity.

The emergence of the God of Time after the Estrangement of Earth from Heaven perfectly underscores the temporal tensions of the earliest Chinese mythmakers, especially when we consider that the god's name, "I Ming," connotes "a sign of sorrow or grief."[7] The identity of the sun as an emblem of time was never more exact; no longer the familiar sun transformed from P'an Ku's left eye nor the gracious god of light and heat, it becomes abstracted into ten suns, each strictly regulated and travelling across the sky through various stations to indicate the periods of the day, and each strictly supervised and fulfilling through rotation the decimal system of dating.

Although there is much disagreement among Chinese scholars about the priority of the appearance of the Myth of the Ten Suns over the decimal division of time used in the Shang period (1766-1123 B.C.), all agree that the names of the ten suns and the decimal points are identical; that the myth of the rotation of the ten suns appeared no later than the Shang period; and that the myth of the simultaneous appearance of ten suns appeared no later than the end of the Spring and Autumn period (722-481 B.C.)[8] At any rate the imaginative basis for the Myth of the Ten Suns can be seen as an increased awareness of the discontinuity of time with its fragmentation into different modes of experience of the ambivalent gods of the four seasons[9] and the Myth of Hsi He and her fleeting chariot forms the background to the anguish of life's ephemerality in the Myth of Titan P'eng Tsu.

---

[5] Cf. T'ang Chün-i, *Chung-kuo wen-hua chih ching-shen chia-chih (The Spiritual Value of Chinese Culture)* (Taipei: Cheng-chung shu-chü, 1972), pp. 22-24.

[6] Cf. Eliade, *Cosmos and History*, p. 75.

[7] Cf. *Tz'u hai (Grand Dictionary)* (Taipei: Chung-hua shu-chü, 1972), pp. 626-29. Also cf. Morohashi, *Dai kanwa jiten*, pp. 1120, 1159 (v. 2).

[8] Cf. Kuan Tung-kuei, "A Study on the Ancient Chinese Myth of the Ten Suns," *Bulletin of the Institute of History and Philology, Academia Sinica*, XXX III (1962, Taipei), pp. 289-317.

[9] Cf. Tu Er-wei, *The Mythological System of the Mountain and Sea Classic*, pp. 1-8 Tu suggests that an ambiguous concept of directions, seasons, and correspondent colours underlies *The Mountain and Sea Classic*. He further notes that the Southern Mountain Classic describes the summer moon, the Western Mountain Classic the autumn moon, the Northern Mountain Classic the winter moon, and the Eastern Mountain Classic the spring moon. With regard to directional colour, the south is usually red, the west white, the north black, and the east green.

In "Life and Immortality in the Mind of Han China", Yü Ying-shih notes that in the bronze inscriptions of the Western Chou period (1122–771 B.C.), pleas for longevity are the most popular inscriptions in the prayers to ancestors or Heaven. But by the Spring and Autumn period there emerged a concept of immortality which differed considerably from its traditional counterpart, and by the end of the Warring States period it had become a widespread cult. Such phrases as "retarding old age," "no death," "ascending to the distant place," "transcending the world," and "becoming immortal," became pervasive literary references as well as inscriptions.[10] The transition from desire for worldly longevity to other-worldly immortality illustrates the continuing psychic stress of the temporal sense. Throughout the early mythic literature we find references to remote countries where death does not exist, residence of the immortals, and the healing herbs and fruits of life on cosmic mountains. The temporal sense revealed in Chinese myth has moved full circle from paradisal enjoyment to profane disillusionment.

b. The Myths of Hou I and Titan K'ua Fu: Myths of the Rebellion Against Time

The records of ancient Chinese myth illustrate that the temporal anxieties of existence naturally gave rise to conflicting responses; in this case either rebellion or reconciliation. The Myth of Titan K'ua Fu belongs to the former. His futile quest to capture the sun and thus put an end to time is a remarkable statement of man's revolt against time's omnipotence. References to the God of Hades, the God of Night, the Abyss Yü as the nadir of the sun's descent, the Great Marshes of Chaos and Life, and the Peach Tree of Immortality suggest a diurnal interpretation of the death and rebirth of the day.[11] Yet the gargantuan thirst of K'ua Fu is a powerful index of the bitterness of mortality realized with the estrangement of Earth from Heaven. Perhaps the transformation of his staff into a peach forest should be seen as the persistence of the will to revolt; in other words a sensual balm to allay the disgust which life's travellers feel toward their certain end.

Man's persistence against time even garners the sympathies and participation of certain gods, as the Myth of Hou I and the Ten Suns illustrates. We have already stressed the importance of the temporal sense in the original Myth of the Ten Suns over and above its usual interpretation as a myth of natural calamity and human strength. Both the implicit contrast between paradisal timelessness

---

[10] Cf. Yü Ying-shih, "Life and Immortality in the Mind of Han China," *Harvard Journal of Asiatic Studies*, 25, (1964–1965), pp. 87–90.

[11] Cf. Wang Hsiao-lien, "K'ua Fu k'ao" (Studies on K'ua Fu), *The Continent Magazine* (Taipei: The Continent Magazine, 1973), 46:2, 1–20.

and consequent mortal time, and the suggestion that the appearance of the multiple suns caused a lack of life-giving water, support a temporal interpretation. Hou I's anger and frustration with the total domination of the sun's (or time's) hold upon mankind evokes a destructive response against time itself, and he is punished severely for his rebellion, however much mankind's feeling may be with the archer god.

c. **The Myth of the Flight to the Moon: A Myth of the Reconciliation with Time**

Since no revolt against time (to elude death) can succeed, it follows that temporal tensions should be moderated by the desire for reconciliation with time, a wish-fulfilment for the timeless state which was man's before the estrangement. The Myth of the Flight to the Moon is of this type. Mount K'un Lun presents a sacred zone,[12] and its spiral road becomes the passage from death to life, from mortality to divinity, from the ephemeral to the eternal. But its accessibility is beyond mortal ability, and Ch'ang O's desire for reconciliation proves too strong and results in her exile from earth. The undercurrent of futility implies that reconciliation with time is impossible within earthly existence. Her banishment to the moon, that archetype of other-life, points out the difficulty of transcending time on the profane soil of earth.

The temporal response of reconciliation is in a sense the typical response of mythmaking, the imaginative identified with the natural; and when the cyclic myth is considered, the spirit of reconciliation is the origin of the cyclic myth. Certainly the cosmic rhythm with its characteristics of periodic creation played an immense part in the elaboration of cyclic concepts. It provided a certain symbolic form and optimistic parallel between the solar, lunar, and seasonal cycles, and the human organic cycle. Through the intense nostalgia for the paradisal archetype and the severe urge to regenerate himself, the primitive mythmaker chose to assimilate or homogenize the human cycle of life and death to the cycle of nature and thus assign equal ontological reality to the experience of time by assuming that the observed laws of nature were one with the unobserved laws of divinity. Susanne Langer makes a similar point: "The eternal regularities of nature, the heavenly motions, the alternations of night and day on earth, the tides of the oceans, are the most obvious metaphors to convey the dawning concepts of life-functions—birth, growth, decadence, and death."[13] And Mircea Eliade concurs: "What is important is that man has felt the need to

---

[12] For K'un Lun as a lunar or cosmic mountain, cf. Tu Er-wei, "The Meaning of K'un Lun Myths," *Contemporary Thought Quarterly* (1961, Taipei), 1:1.

[13] Langer, *Philosophy in a New Key*, p. 164.

reproduce the cosmogony in his constructions, whatever be their nature; that this reproduction made him contemporary with the mystical moment of the beginning of the world and that he felt the need of returning to the moment, as often as possible, in order to regenerate himself."[14]

d. The Myth of the Divine Administration: A Myth of the Eternal Becoming

Beginning with the myths of creation and heaven, and the consequent estrangement, we have seen the Chinese temporal sense develop through rebellion to reconciliation and finally acceptance of the natural cycle as the metaphor for the regeneration of time and mankind. As the basis of a worldview which has existed for three thousand years, the evolving cyclic myth must not be seen as one myth (as in The Flight to the Moon) but as a pattern within Chinese mythography which despite its malleability controlled the development of future thought. As a pattern of mythopoesis the cyclic myth engendered a fantastic surge of abstraction, once the resolution of temporal tensions became possible through the union of human and natural cycles. The ancient Chinese classics are testimony to the incredible systematizing of correspondences to ensure the identity of man with nature. The descriptions and stories concerning gods of directions, elements, and seasons, scattered through such classics as *The Mountain and Sea Classic*, "the Monthly Observances" in *The Spring and Autumn Annals of Lü*, "The Monthly Ordinances" in *The Book of Rites* and *Huai-nan tzu*, are remarkable records of this tendency to rationalize a homogeneity between biological and cosmic rhythms.[15] One of the major concerns was the abstraction of seasonal rotation into periodic regeneration by

---

[14] Eliade, *Cosmos and History*, pp. 76–77.

[15] *The Spring and Autumn Annals of Lü* (*Lü-shih ch'un-ch'iu*) was compiled by scholars assembled and patronized by Lü Pu-wei, the Prime Minister of the Ch'in dynasty, and was published in 238 B.C. The contents of its first twelve sections (the Monthly Observances or Yüeh ling) are descriptions of royal and baronial beliefs and practices. Its original production may be ascribed to court diviners and scribes, for its ritual was an essential part of their practice, and the welfare of the state depended upon their correct interpretation of its ordinances. *The Book of Rites* (*Li chi*), a compilation of royal ceremonies and duties preserved by the court writers of the Chou dynasty, was lost after the Burning of Books and was revised circa 100 B.C. The title of section IV of the book is also known as "the Yüeh-ling" and many postulate a common source for both. Whether the "Monthly Observances" present an ideal or factual account of the ordinances of government and ritual has been a subject of debate among Chinese scholars. In either case the principles underlying these observances have been embodied in the theory and practice of kingship until almost the present day, and as such, the work is valuable within the framework of this paper.

38 A Comparative Study of Chinese and Western Cyclic Myths

means of an identification of the gods of directions with the gods of elements and seasons. The culmination of these rationalizing labours may be seen in surviving fragments of the Myth of Divine Administration.

Through study of inscriptions on early oracle bones (circa 1300 B.C.) it is evident that concepts of the gods of direction were relatively undeveloped—that is no specific divine character had been assigned to each god. By the end of the fifth century B.C. these gods had developed the natures of the Chinese geographic directions and the virtues of the five elements.[16] These rudiments of the Myth of Divine Administration were further systematized, as the gods of the seasons became the divine assistants of the original five gods of direction. In this manner the regenerative cycle of nature became assimilated more completely within the sphere of human experience.

Originally thought to be a "divine comedy" of man's affiliation with natural periodicity, the Myth of Divine Administration survives only in occasional fragments in the ancient classics. Unfortunately what remains are thematic summaries rather than dramatic plots or records of mythopoeic incidents. However, etymological studies enable us to recapture much of the vitality of this major step in the development of the cyclic myth and the chart at the end of the essay outlines the pattern of analogy and correspondence which is central to the myth's importance.

Chuan Hsü, the Sovereign God of the North, denotes "a cautious refraining," and implies the hibernation of earth after a cold dew. The Winter God's sacred animal is a black tortoise, symbol of the mysterious state whereby the earth retreats beneath the hoarfrost of the north wind.[17] Yü Ch'iang, the assistant, is the son of the Pu Chou Wind which dwells in the obscure snow-covered northwest of the universe and is always associated with death. Despite this grim aspect, the connotation of "Yü Ch'iang" is "at ease with power, in harmony with vigour, or matched with vitality." Thus Yü Ch'iang's image as the God of the North Sea or the God of Water is a fish capable of transforming itself into a gigantic phoenix which flies from the frigid north, bringing the water of life for the spring.[18] Yü Ch'iang is also called "Yüan Meng" (denoting "origin in occult darkness or chaos") or "Hsüan Meng" (denoting "latent profundity and perseverance").[19] To summarize, Chuan Hsü and his

---

[16] Cf. Chou He, *Ch'un-ch'iu chi-li k'ao-pien* (*Investigation of Rituals and Ceremonies of the Spring and Autumn Period*) (Taipei: Chia-hsin wen-hua chi-chin-hui, 1970), pp. 17–18.

[17] Cf. *Tz'u hai*, p. 3178; Morohashi, *Dai kanwa jiten*, p. 291 (v. 12).

[18] Cf. *Tz'u hai*, pp. 1076, 1088, 1091; *Dai kanwa jiten*, pp. 771 (v. 4), 517 (v. 8).

[19] Cf. *Tz'u hai*, pp. 284, 360, 1904; *Dai kanwa jiten*, pp. 973 (v. 1), 130 (v. 2), 765 (v. 7), 775 (v. 7).

Aspects of Cyclic Myth in Chinese Mythology 39

assistant, Yü Ch'iang, connote the latent vitality of darkness and chaos, the potential for growth hidden in the deepest snows of winter.

Through examination of the relationship between other gods and their assistants, the connotations of archaic names, and the symbolic reasoning behind certain sacred animals, the pattern of the Myth of Divine Administration emerges. However fragmentary the original sources, this myth embodies that tendency to abstract the human to the divine through the intermediary example of natural cycles.

The hierophany of the gods of directions vividly illustrates the nascent cyclic myth: birth, growth, completion, hibernation (abstention and latency), and death, followed by the glorious rebirth. The divine administration—the division of experience into symmetrical quadrants—demonstrates an urgent will beneath the abstraction. It is an attempt under severe temporal tension to identify the dark side of death or the ephemerality of being with diurnal change, the lunar cycle, and the seasonal rotation which promises life through death. It is also the hope of transcendence from a life of fleeting existence to a life of eternal becoming. In creating a strong affiliation of the gods of directions with the gods of elements and seasons, a periodic regeneration is established, and a renewal of life is obtained by man as a part of nature.

B. ASPECTS OF CYCLIC MYTH IN CHINESE RITUALS AND EARLY BELIEFS

1. Ritual: A Mode of Actualization

Before turning to the development of the cyclic myth in Chinese ritual, it is imperative to dispense with a maze of definition, for if Western critics have debated "myth" into a labyrinth, then surely the central Minotaur must be the relation between "myth" and "ritual." At the outset it must be reiterated that prescribing the primacy of one or the other is not really to the point of this study; it is preferable to be descriptive rather than prescriptive, and follow Clyde Kluckholm's lead: "This relationship is not one of the primacy of either case, but that of an intricate mutual interdependence, differently structured in different cultures."[1] With respect to the cyclic myth, Chinese ritual practice has a common psychological basis, and offers a formalized statement or symbolic dramatization of the same needs.

The problem is best summarized with reference to Philip Wheelwright: "As the primitive participates in nature, alternations of movement and rest as there may be are soon accentuated and dramatized by ritual so that the human transition may blend with that of the cosmos.... Nature, in riteopoeic perspective as well as mythopoeic perspective, is cyclical; it exhibits vitally periodic becoming."[2]

2. Totemism

An investigation of the cyclic basis of early Chinese ritual begins with the most primitive forms of totemism, a phenomenon believed non-existent in China by Western social scientists until several decades ago. The supposed absence of totemism in ancient China would seem to present grave problems to those theorists who believe that the "totemic-era" is an inevitable cultural period of the history of mankind. Fortunately the pioneering works of Li Chi and Huang Wen-shan have unveiled the totemistic practices of Chinese society during the Hsia dynasty (2205–1766 B.C.) and during the Lower Palaeolithic and Neolithic Ages.[3]

---

[1] Clyde Kluckholm, "Myth and Ritual: A General Theory," in John B. Vickery, ed., *Myth and Literature: Contemporary Theory and Practice*, (Lincoln: University of Nebraska Press, 1966), p. 39.

[2] Philip Wheelwright, "Notes on Mythopoeia," in Vickery, pp. 60, 64.

[3] Cf. Li Chi, *Beginnings of Chinese Civilizations* (Seattle: University of Washington Press, 1957), pp. 20–21. Also cf. Huang Wen-shan, "Totemism and the Origin of Chinese Philosophy," in *Bulletin of the Institute of Ethnology, Academia Sinica*, 9, (Taipei, 1960), p. 52.

In confining our discussion to aspects of totemism in the formation of the Chinese cyclic myth, we are indebted to Liu Chieh's systematic and scrupulous study, *History of the Migrations of Ancient Chinese Gentes*, which asserts that the lizard, among the three earliest totemic animals, and the sun and the moon were totems of the clans of the Archer God Hou I's offspring; and that the decimal names for the ten suns and later that of their corresponding animals were also the names of various phratries in the Hsia dynasty.[4] The original character for "lizard" is a paradoxical combination of the sun and moon which etymologically connotes the lizard's protective coloration, that is, to change as or with the light from the sun and moon.[5] At any rate the significance of this totem resides in the virtue of adaptation: the persistence of the ever-rising sun and the eternally changing moon.[6]

Here, present in one of the earliest known Chinese totems is the restless urge to alleviate that temporal tension which gave rise to the myth of the Archer God in the first place. Again the desire for harmony with cyclic nature, the nostalgia for periodic regeneration, is elucidated and emphasized.

3. The Sacrifice to Heaven and Earth

If the most archaic classics are any indication, the ancient Chinese were extremely ritualistic, and numerous are the records of strict and intricate rituals of daily, monthly, seasonal, and yearly sacrifices. Among these the sacrifices to Heaven and Earth, to the four directions, and to the cosmic mountain are most important either as reflections of or contributions to the evolution of the cyclic myth. But before any further study, what must be stressed here are the sun and the moon as the images of Heaven and Earth. From time immemorial it has been believed that the sun, born in the morning in the east at the spring equinox, is the essence of Heaven which in every aspect is luminous, active, and diffusive; while the moon, born in the evening in the west at the autumn equinox, is the essence of Earth which in all aspects is nebulous, passive, and accommodating.[7]

The Sacrifice to Heaven and Earth is first recorded in the inscriptions on oracle bones and is called "Chiao," literally "the Suburban Sacrifice." As the greatest sacrifice of the year, it was offered by the emperor as the Son of

---

[4] Cf. Liu Chieh, *History of the Migrations of Ancient Chinese Gentes* (Taipei: Cheng-chung shu-chü, 1971), pp. 50–98.

[5] Cf. Liu Chieh, *History of the Migrations of Ancient Chinese Gentes*, p. 88. Also cf. *Etymological Dictionary (Shuo-wen chieh-tzu)* cited by Liu Chieh, p. 53.

[6] Cf. stanza from *The Book of Odes*, cited by Liu Chieh, p. 88.

[7] From various classics cited by John Er-wei Tu, *The Religious System of Ancient China* (Taipei: Hua-ming shu-chü, 1951), pp. 126, 155.

Heaven and the Representative of Earth. The rite of Chiao is divided into two sacrifices: the Earth Sacrifice must be held on a square mound on the summer solstice in a southern marsh; the Sacrifice to Heaven must occur on a round mound on the winter solstice on a northern hill. While the square and the round mounds are supposedly imitative of Earth and Heaven respectively, it is the temporal considerations which are most revealing of the cyclic concept.

The summer solstice was believed to be the day the moon (as the essence of Earth) was at its extreme southerly position, the point of revival of the powers of dormancy, quiescence, and abstention. The winter solstice was the day the sun was most northerly, and the point of revival of the powers of vitality, exuberance, and diffusion. Ancient custom dictated that nothing should be attempted which might hinder the solar or lunar return on these two days. Thus on the summer solstice we read: "No fires must be lit in the southern part of the house lest the heat be over-encouraged. Doors and gates must be closed to encourage the life-force and the free flow of the seasonal influence. Men of position or rank must keep vigil and fast. They must remain secluded in their house, avoid violent exertion, abstain from music and the beautiful, avoid sexual indulgence...."[8] Strict obedience of these customs would ensure the seasonal victory of the forces of decay and darkness and preserve the vegetative cycle.

Regarding the ascendancy of the powers of growth and light during the winter solstice, the following customs were observed: "In the eleventh hour responsible officials are commanded to take care that nothing covered be thrown open, and that there should be no calling up of the masses. No dwellings should be thrown open: no digging should be done or the heat of the earth would escape, or be stirred to unseasonable activity."[9] For the Western mind the rites of Chiao must seem incredibly naive, even foolish, but they occupy a central position in the development of the cyclic myth. Here for the first time we see that through rituals corresponding to the cosmic rhythm, the emperor as the anointed superintendent of the world could ensure harmony between the human and natural worlds. The Chiao is a graphic example of the consuming desire of the early Chinese mind for periodic regeneration, and the annulment of time.

The round mound of the first rites of Chiao was the prototype of the final Altar to Heaven in Peking, and the ritual the original pattern of the emperor's role in Chinese life for succeeding centuries. The rite was last performed in

---

[8] For "Yüeh ling" in *Li chi*, cf. Wang Meng-ou, *Li chi chin-chu chin-i* (Taipei: Shang-wu yin-shu-kuan, 1971), pp. 214–19. Also cf. William E. Soothill, *The Hall of Light* (London: Lutterworth Press, 1951), pp. 42–43.

[9] Cf. Wang Meng-ou, *Li chi chin-chu chin-i*, pp. 232–41. Also cf. Soothill, *The Hall of Light*, pp. 48–49.

1935 under Japanese occupation and, despite the use of radio broadcast, recaptured much of its original power.[10]

4. The Sacrifice to the Four Directions

Aside from the emperor's responsibility to regulate the myriads of living things along the righteous course of the universe and thereby maintain harmony between man, Heaven and Earth, the four directions received no less attention in ancient rites. According to "Interpretation of Rites" in *The Book of Rites*, sacrifices should be offered at the beginning of each season to the corresponding gods of directions and seasons to encourage the harmonious course of each season. Six jade offerings used in these ceremonies are vivid reminders of the Myth of Divine Administration and the cyclic basis of the rite: Pi, a round green tablet in imitation of the perfect cycle of the cosmic rhythm, was to be offered in homage to Heaven. Ts'ung, a square yellow tablet, symbolizing the great ground, was to be offered in homage to the God of Earth. Kuei, a green equilateral-triangle tablet in imitation of new buds and sprouts of the spring, was to be offered in homage to the Sovereign God of the East and the God of the Spring. Chang, a red acute-right-angle triangle tablet, symbolizing the half-death of the vitality of all summer things, was to be offered in homage to the Sovereign God of the South and the God of the Summer. Hu, a white tiger-shaped tablet, symbolizing the prey of autumn's severity, was to be offered in homage to the Sovereign God of the West and the God of the Autumn. And finally, Huang, a black semi-circular tablet, symbolizing the half-death or death-like state of dormancy and hibernation in the winter, was to be offered in homage to the Sovereign God of the North and the God of the Winter.[11] Indeed, in this ancient rite, the psyche to blend the human transition with the periodic becoming of the cosmos is strikingly evident.

---

[10] Cf. Soothill, *The Hall of Light*, p. 189. Hsüan T'ung, the dethroned emperor of the Ch'ing dynasty, under Japanese pressure returned to his ancestral Manchuria to found a new dynasty. A new tricentric Altar designed in imitation of the historic Peking altar was erected and the rites performed as Soothill describes: "At the winter solstice 1935, in the darkness before dawn he came in the rightful robes, following the ancient ritual, if with diminished glory, knelt under the frosty stars, facing the north; and by radio the world listened to his thin sharp voice as it rose and fell while he offered himself to the ancestors and the powers above and of nature—the Shang Ti, and cried aloud the ancient prayers for suitable seasons for his new-old people. His sacrificial pyres were scanty compared with the past, his retinue small; but some of his supporters may have hoped that through the renewed rites, the 'kingly way' might come back to earth, and potency and virtue flow into the slight robed figure looking toward the north star, bearing himself with a pathetic dignity after the fashion of the sovereigns of a long, long past."

[11] Cf. "Ta tsung-po" and annotation by Cheng Hsüan, cited by John Er-wei Tu, *The Religious System of Ancient China*, p. 149.

5. The Sacrifice to the Cosmic Mountain

The Feng Shan Sacrifice was a mysterious ancient rite which had been practised up until Sung dynasty (960–1279 A.D.) Almost every emperor after enthronement would go to Mount T'ai to offer thanksgiving sacrifices to Heaven and Earth. The Sacrifice to Earth was held at the foot of Mount T'ai and the emperor would symbolically earth up the hill to reinforce the foundation or the root of the mountain; while the Sacrifice to Heaven was held on the top of Mount T'ai and the emperor would symbolically build a mound to elevate the height of the mountain. As the Son of Heaven and its Representative on Earth, the emperor was originally to return and make homage to Mount K'un Lun, the cosmic mountain upon which centred the myths of passageways to Heaven, the myths of divine abodes of the Supreme Being and Supernal Mother Goddesses, and the myths of the Trees of Life in various gardens. However, the mythic location and inaccessibility of Mount K'un Lun prompted the substitution of Mount T'ai in its place. At any rate, what is more important is the mythic undercurrent: this ancient rite is a return to the origin through which immortality might be obtained. It recalls the Myth of Archer God Hou I's Return to Mount K'un Lun and the Myth of Goddess Ch'ang O's Flight to the Moon. The dramatization of the emperor's wish for the eternal return by elevating the mountain top so it would perhaps reach Heaven and by banking up the foot of the hill lest it would topple is most romantic and touching.[12]

To review the elaboration of the cyclic myth in these earliest Chinese rituals is to underscore the importance of temporal tensions in the creation of these rites. The most archaic sun-moon totemism may be viewed as a nostalgia for periodic regeneration; the sacrifices to Heaven and Earth and the Four Directions are attempts to harmonize human life with the cosmic rhythm of eternal return, while the thanksgiving sacrifice to the cosmic mountain typifies that desire to return to the origin, or gain immortality. As the early myths became codified and abstracted into the complex Myth of Divine Administration, the riteopoeic aspect generated abstractions of ritual culminating in the great institution "Ming t'ang," the Hall of Light.

6. The Institution of the Hall of Light

Any investigation of ancient Chinese ritual invariably leads to the institution of "Ming t'ang," literally the Hall of Light, from which sprang Chinese astronomy, cosmology, religion, government, agriculture, and ethics. While we

---

[12] See "Feng jen" in *Chou li*, "Pao fu chu" in *Ta Tai li*, and "Chiao szu chih" in *Han shu*, cited by Tu Er-wei, *The Religious System of Ancient China*, pp. 129–47.

have suggested the conceptual framework which gave rise to the institution, its actual origin is shrouded in conjecture. The denotation of "ming" first suggests an astronomical origin or association, for "ming" is composed of two pictographs for the sun and the moon, and therefore indicates that which is produced by the two luminaries. Most Chinese classics suggest the origin of Ming t'ang to be no more than the thatched hut of the tribal magician-astronomer, or perhaps the secluded sanctum used by the sage-ruler for astronomical observation.[13] Tradition assigns the building of the first Ming t'ang to Shen Nung, the first legendary ruler and Patriarch of Agriculture (probably the euhemerized figure of the great Sun-god of the South), in the shape of the Pa Kua, the octagonal form of astronomical changes invented by Fu Hsi, the Sovereign God of the East. Whatever its origin, it is assured that during the expansion from a tribal to feudal system, the thatched hut became first five rooms, and then in the Chou dynasty, nine halls surmounted by an upper circular story set on the square central hall.[14]

---

[13] Cf. Soothill, *The Hall of Light*, pp. 67-68. Our discussion of Ming T'ang is almost entirely indebted to William E. Soothill's remarkably detailed investigations in *The Hall of Light*. It is indispensable for students of early Chinese history and philosophy.

[14] The arrangement of the rooms has been a matter for controversy among scholars of Chinese ritual, and the chart here shows a version with 12 halls, modified from five room construction. Also cf. Soothill, *The Hall of Light*, pp. 88-89.

46 *A Comparative Study of Chinese and Western Cyclic Myths*

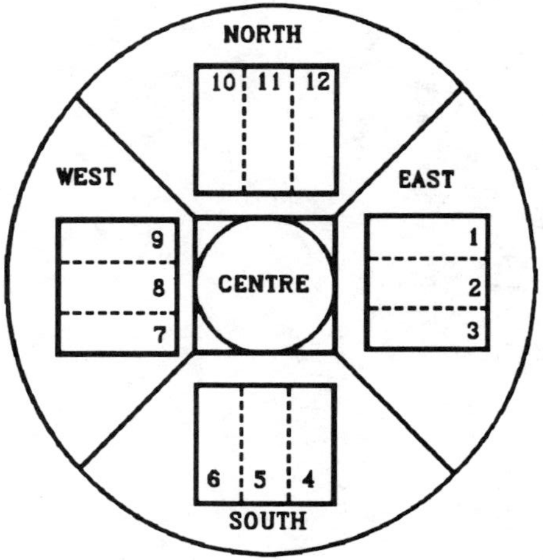

East/Spring/Green/Wood
Hall 1/ left division/ February
Hall 2/ Hall of Azure Sun/ March
Hall 3/ right division/ April
Requirement: all chariot, horse, banner, garment, and jade ornament in green.

North/Winter/Black/Water
Hall 10/ left division/ November
Hall 11/ Somber Hall/ December
Hall 12/ right division/ January
Requirement: all chariot, horse, banner, garment, and jade ornament in black.

South/Summer/Red/Fire
Hall 4/ left division/ May
Hall 5/ Hall of Light/ June
Hall 6/ right division/ July
Requirement: all chariot, horse, banner, garment, and jade ornament in red.

West/Autumn/White/Metal
Hall 7/ left division/ August
Hall 8/ Hall of Assembly/ September
Hall 9/ right division/ October
Requirement: all chariot, horse, banner, garment, and jade ornament in whate.

Centre/Midsummer/Yellow/Soil
(The day of the Saturn)
(between July and August)
Requirement: all chariot, horse, banner, garment, and jade ornament in yellow.

The sundial structure of Ming t'ang represents both the terrestrial and celestial centre of the Chinese world. The upper story or astronomical observatory was accorded the name "K'un Lun" (from the cosmic mountain), while the square below was said to be symbolic of Earth in opposition to the circular Heaven above. Understood as a directional and seasonal scheme, the five or nine halls provide a clockwise progression of the monthly sacrifices from the beginning to the end of the year. Records show that "it was the duty of the ruler at each new moon to prepare himself according to various rules, by ablutions, by eating special foods varying with the seasons, by wearing the garments of the seasons, and performing various other duties required of him, to offer the monthly mimetic sacrifices."[15] An improper sacrifice would not merely be inefficacious, but inauspicious to the point of disjointing the course of nature. Therefore the ruler, or emperor, should act in accord with the scheme of progression and "maintain it in world-wide exactitude in order to link up the ways of the three powers of Heaven, Earth, and Man, and to extend it throughout the seasons, for the three powers and the four seasons form the seven essentials, from which spring the fulfilment of the nature of things in general and which, by aiding their development and nurture, complete that which has already been ordained."[16] In practice the emperor would progressively occupy a room facing the direction indicated by the month and perform the required cosmological, astronomical, social, and agricultural rituals and ceremonies.

With the breakup of the feudal system of the Chou dynasty, Ming t'ang fell into decline and its functions as an administrative centre were dispersed through the empire. Yet it remained the Son of Heaven's greatest temple, as the need for the nation to be linked in harmony with the realm of nature still remained, and for this purpose the ruler was still the one essential nexus. In all operations the animated world still depended upon his cooperation in the threefold union of Man, Earth, and Heaven; the Ming t'ang remained the nation's powerhouse for its effects were political and ethical which, for the ancient Chinese, were in essence religious.

The institution of Ming t'ang suggests the intense symbiosis between man and cosmos which is a hallmark of Chinese culture. William E. Soothill's explanation parallels our explanation of Ming t'ang as a crucial manifestation of the evolving cyclic myth: "This unitary and symmetrical system of celestial palaces is but a conviction of the existence of an universal law, a revelation of a cyclic concept of the universe, an awareness of the destiny or necessity of man's participation in the due course of universal harmony, and hence a manifestation of an age-old dream of manipulation over it through the sage-ruler

---

[15] Cf. Wang Meng-ou, *Li chi chin-chu chin-i*, pp. 201–43.
[16] Soothill, *The Hall of Light*, p. 70.

or the emperor's knowledge of, and therefore the power over, the cosmic order with its annual cycle of months and seasons."[17] As the basis of the Chinese calendar and cosmology, the institution of Ming t'ang played a major role in the development of tetra-symmetrical and penta-cyclical concepts so important in the evolution of Chinese philosophy—a philosophy which would in return contribute much to the design of the cyclic pattern of man in the cosmos.

The progressive anthropocentrism of Chinese ritual revealed in the movement from the earliest totemism, through the rites of Chiao, and finally in the developed institution of Ming t'ang is intrinsically linked to increasing abstractions of natural periodicity, for only as man is in harmony with cyclic nature does he approach the divine. The temporal tensions motivating this progression are also visible in the shift from primitive ancestor-worship (as a means of communing with divinity) to the concept of "Te" as the essence of the Mandate of Heaven, a crucial development in maintaining the legitimacy of Ming t'ang.

7. The Ancestor Worship Under a Personal Heaven

Inscriptions on more than one hundred thousand fragments of oracle bones bear powerful testimony to the potency of ancestor worship in the Shang dynasty (1766–1123 B.C.). The vast majority of these are records of the Shang rulers' pleas, requests and consultations regarding affairs of harvest, warfare, calamity and general social welfare. What is most interesting in these invocations is the utter necessity of intercession by the great ancestors in communications with the Supreme Being, "Ti." It seems that accessibility to the divine world was virtually impossible except for the great ancestor or forerunners of the dynasty. Ancestor deification is perhaps traced to the Shang belief that death was a natural return to the origin and mysterious state of increased spiritual potency; the semi-divine dynasty founder thus dwelt "on high," in close association with the Supreme Being, and was able to intercede on behalf of the consulting ruler. The people of the Shang were more inclined to synthesize the powers of the Supreme Being and the deified ancestor, rather than distinguish between them. This homogenization is confirmed by Wang Kuo-wei who concludes that "Ti" and the great ancestor were but two faces of the same identity.[18] Etymological study reinforces this conclusion: the pictographic character for "ancestor" takes the form of a phallus, while that of "the Supreme Being" and "the sacrifices to great ancestors of the dynasty" appears as a flower with ovary. Both characters are symbols of origin and regeneration; indeed the desire to return to the origin,

---

[17] Soothill, *The Hall of Light*, p. 112.
[18] Cf. Kuo Ting-t'ang, *The Development of the Concept of Heaven in the Pre-Ch'in Period* (Shanghai: Shang-wu yin-shu-kuan, 1936) pp. 11–16.

to unite with the regenerative powers of the divine world, may be seen as the psychological and ethical basis of the Ancestor Worship.[19]

8. The Mandate of an Impersonal Heaven

With the fall of the Shang dynasty, the Supreme Being "Ti" was soon substituted by the "Omnipotence of T'ien" of the Chou rulers. This "T'ien" of the Chou was a far less "personal" omnipotence and associated with the "omnigoodness" and "omnijustice" of the universe. The rationalization of the concept of the Supreme Being was in part due to the divine justification sought by the Chou rulers for their revolution. Thus the Chou claimed a "Mandate of Heaven," which favoured no one but the virtuous, appointed no eternal mandatary, and promised no eternal blessing. The concept of "Te" or virtue is central to the "impersonalization of Heaven" implied in the Mandate of Heaven.

"Te" must be understood not only in the English sense of virtue, but also as an expression of natural harmony; inwardly a spontaneity with this harmony, and outwardly as the sustaining constancy or the irresistible potency of the threefold unity of Heaven, Earth, and Man. It was believed that Heaven would withdraw its mandate from an unworthy occupant of the throne, who stood between the transmission of "virtue" between the celestial realm and the people. The concept of "Te" then served to undercut the transcendency of the celestial and innovate the immanent counterpoise of the terrestrial in the cosmic harmony.

Through the Influence of "Te," the Mandate of Heaven became more rational and humanistic, and finally became identified with the "collective will" of the people upon the exile of the tyrannic King Li by the people of the Chou.[20] Through such vagaries of history, the "Mandate" became less a justification of kingship and more an excuse, yet the increasing need for amelioration of temporal tensions through union with the natural cycle remained, and eventually stimulated a more cosmogonic concept of "virtue," the Tao, or the most basic stratum of Chinese philosophy.

9. Tao as a Totalistic Concept of an Ideal and Universal Order

As one of the central ideas of Chinese life, Tao has been the subject of intense scrutiny by Chinese and other scholars. Some suggest that "Tao" was originally a totemistic principle comparable to "mana" and Tu Er-wei refines this concept

---

[19] Cf. Wei Cheng-t'ung, *A Critical Approach to the Chinese Culture* (Taipei: Shui-niu ch'u-pan-she, 1969), p. 79. Also cf. *Ch'un-ch'iu chi-li k'ao-pien*, pp. 138–39.

[20] Cf. Yeh Yü-Lin, annotated, *Kuo yü* (Taipei: Shang-wu yin-shu-kuan, 1970), pp. 10–13. Also cf. Wei Cheng-t'ung, *A Critical Approach to the Chinese Philosophy and Thought* (Taipei: Shui-niu ch'u-pan-she, 1968), p. 17; Kuo Ting-t'ang, p. 33.

to the identity of "Tao" as the exact name and image for the impersonal power of the lunar cycle.[21] Both suggestions are implied in an early meaning of Tao as the way of nature or the sustaining potency of the course of nature. By the later Chou period it had become all-embracing and yet, somehow transcending the ancient symbols, finally bore a great resemblance to the logos of Hellenistic and Greco-Roman philosophers. It became a totalistic concept of an ideal and universal order and harmony, and an ultimate assertion of man's oneness with nature. However, its philosophical elaboration is perhaps better discussed within the framework of that seminal Chinese work, *The Book of Changes*.

10. Summary: From Peasant Omen to a Microcosm

It is indeed fitting to summarize the importance of the cyclic myth in early Chinese ritual and belief with a brief study of *The Book of Changes* (*I Ching*), a work described by Richard Wilhelm as a foundation of Chinese life: "Nearly all that is greatest and most significant in three thousand years of Chinese cultural history has either taken its inspiration from this book, or has exerted an influence on the interpretation of its text."[22] Fortunately for our purposes, the Western sinologists have the pioneering works of James Legge and Hellmut Wilhelm readily available and numerous secondary studies also translated into English; we may thus restrict ourselves to the influence of the cyclic myth in the formation of *The Book of Changes*.

Concerning the earliest history of *The Book of Changes*, it was perhaps at first an unorganized collection of peasant omen-texts, then during the Hsia dynasty these documents were combined with divination practices. Towards the beginning of the Chou dynasty it had assumed its present form of the Eight Trigrams (the Pa kua) and probably the permutation of it into the sixty-four hexagrams. Despite the impossibility of establishing an accurate chronology, we can deduce from the development of *The Book of Changes* the importance of temporal tensions and the desire for periodic regeneration so typical of the mythology of temporal awareness outlined earlier.

The efficacy of divination is a central concept in Chinese history, and one that extends into remotest antiquity. Originating in agricultural auguries, the practice of divination has always implied a communication between microcosm and macrocosm, a belief in the parallelism between what is without and what is within us, that the course of one may also be meaningful for the other. While the processes of natural change are too varied for immediate perception, the

---

[21] Cf. Huang Wen-shan, "Totemism and the Origin of Chinese Philosophy", pp. 54–56. Also cf. Tu Er-wei, *The Religious System of Ancient China*, pp. 1–15.

[22] Richard Wilhelm, *The I Ching: Book of Changes*, trans., Cary F. Baynes (Princeton: Princeton University Press, 3rd ed., 1967), p. xxvii.

direction of change may be perceived through the use of sacred objects which act as the medium between human action and natural law. With the "Myth of Estrangement of Earth from Heaven," divination must also be seen as the desire for reconciliation between human and natural time; one essentially linear, the other cyclic.

Connections between divination practices and what would become *The Book of Changes* were established during the Shang dynasty by the codification of their eight natural deities (Heaven, Earth, mountain, marsh, thunder, fire, water, and wood) into eight lineal configurations, the Pa kua. Divination of the Pa kua was accomplished through the use of the tortoise-shell and the arrangement of stalks of milfoil. The tortoise was the sacred animal of the north-winter, a hibernal animal capable of seemingly occult death and rebirth, while the milfoil was regarded as a sacred plant bearing three hundred stems every thousand years and connected with the virtue of roundness or perfection. This natural symbolism further underscores the cyclic basis of divination; in both media we can discern the importance of regeneration and cyclic renewal.

The origins of the Pa kua are clouded by the lack of historical data. Chinese tradition attributes the invention of the eight images to Fu Hsi, Sovereign God of the East, and the first euhemerized emperor of the third millennium B.C. Fu Hsi allegedly contemplated the forms and patterns exhibited in sky, earth and self, and devised the eight lineal figures of three lines each to demonstrate the intelligent operations of nature and classify the qualities of a myriad things. Fu Hsi is also credited with ordering the Pa kua into the "Primal Arrangement" which is basically a symbolic statement of the eternal and balanced completeness within which the process of dynamic balance operates. In the Primal Arrangement is found the interconnection of element, season, and direction which typifies the abstract systematization of the Myth of Divine Administration, the Chiao sacrifice, and the institution of Ming t'ang.[23]

The roots of *The Book of Changes* are further grounded in natural cycles when the importance of celestial objects is considered. "I" or "Change" is written as a pictograph of the sun placed over the moon, and as in the heavens, the sun is replaced by the moon, and the moon again by the sun, so "Chang"

---

[23] K. A. Dhiegh's *The Eleventh Wing* (New York: Dell Publishing, 1973, pp. 81–92) offers an interesting analysis of the symbolic indications contained in the Primal Arrangement. He notes that: "Man's joyous temperament (Tui) is situated in the southeast octagonal wedge between Li and Ch'ien. Li is the natural element of fire. It represents light, clarity, sun, eye, perception and understanding. It is set in the east, a symbol of rising, beginning, a starting point from where there is ascendance. Above, in the south, is Ch'ien (The Originating). Ch'ien is the source of all phenomena, projective, the root of power.... An inference: Man experiences the joyous state when he perceives without a doubt that he functions as a rising expression of heaven's power, and with this conviction he has merged into the will of heaven" (pp. 87–88).

always proceeds in the phenomena of nature and the experiences of society. Richard Wilhelm has a better annotation to this point: "Owing to changes of the sun, moon and stars, phenomena take form in the heavens. These come into being on earth, in accordance with identical laws. Therefore the processes on earth—blossom and fruit, growth and decay—can be calculated if we know the laws of time."[24] The premise for such a speculation is a belief in the concept of constancy evoked by the orbit of heavenly bodies, and the relationship between the polarities of heaven and earth, sun and moon, and the yin and yang principles. Obviously these celestial phenomena have always been present and ready for abstraction. In due course, images were recognized, attributes assigned, and concepts formed. And it is generally believed that the Shang were the first to develop the concepts of polarity between heaven and earth, masculine and feminine, and yin and yang. The following diagram illustrates the primal arrangement of the Pa kua in pairs of antitheses.

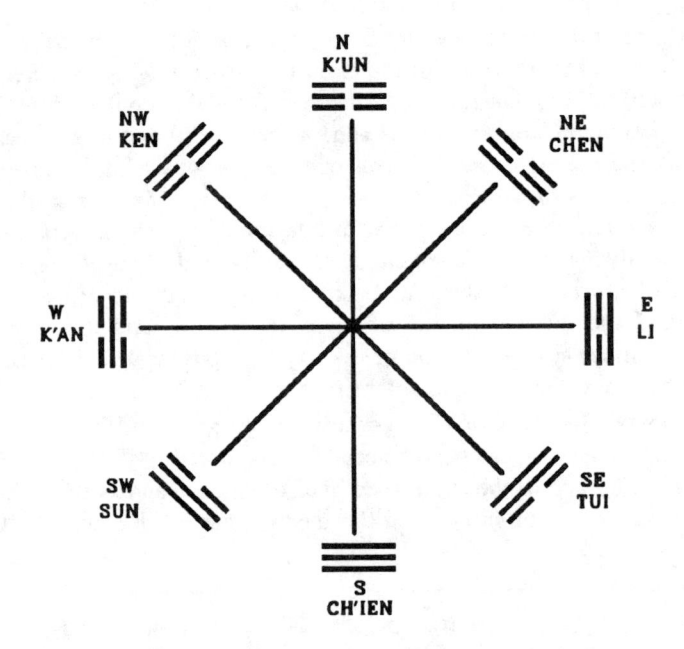

The Primal Arrangement of the Pa Kua in
Pairs of Antitheses

---

[24] Richard Wilhelm, *The I Ching: Book of Changes*, p. 283.

The eight trigrams of the Pa kua were of course not immune to the rationalizing complexities of the early Chinese mind. The tensions which compelled the abstracting correspondences between man, nature, and the divine in previous schemata (Ming t'ang, etc.) are also behind the evolving Pa kua. The accretion of correspondences was historically reinforced by the fall of the House of Shang, for royal diviners were exiled and forced to earn a living from divining among common citizens. To meet the various applications of "terrestrial" life, many more images and attributes were added to the Pa kua. The following chart is offered in illustration of the increasing complexity of symbols and attributes.

In order for the cyclic nature of change to be intelligible in such a linear configuration, it must be pointed out that the two cardinal principles of all existence are symbolized by the trigrams: Ch'ien (The Creative) and K'un (The Receptive). The two principles are united by a relation based on homogeneity: they do not combat but complement each other. The difference in level creates a potential by virtue of which movement and living expression of energy become possible. Out of this potential there stirs the active force Chen (The Arousing), symbolized by thunder. This electrical force forms centres of activation which are discharged as lightning. This dependent force is Li (The Clinging). Now the movement shifts and thunder's opposite, the wind, sets in. This gentle, flexible but penetrating force is called Sun (The Penetrating). The wind is followed by rain. This enveloping, dangerous and difficult force is called K'an (The Abysmal). Then there is a new shift as the trigrams Li and K'an (formerly acting in their secondary forms as lightning and rain) now appear in their primary forms as the sun and the moon which in turn cause heat and cold. When the sun reaches the zenith there is heat. This joyful, satisfied and complacent force is call Tui (The Joyous). When the moon reaches the zenith there is cold. This resting, stubborn and unmoving force is represented by Ken (The Mountain). The tendency of keeping still in the trigram Ken is of mysterious significance for here in the seed, in the deep-hidden stillness, the end of everything is joined to a new beginning.[25] We have then in the Pa kua the roots of the first Chinese metaphysical speculation which proclaims that from Tao emerged the great primordial One; the One evolved into the Two polarities of Yin and Yang; the Two produced the Four seasonal phenomena, or heavenly spheres, or phases of being; and the Four in turn produced the Eight trigrams which are symbols of bio-cosmic transformation. Above all it must be emphasized that the Pa kua are not representations of things as such, but of their tendencies in movement, natural processes or phases corresponding with their inherent attributes; and as such, the Pa kua, through divination, offer man communication with regenerative nature and release from temporal limitations.

---

[25] Cf. Richard Wilhelm, *The I Ching: Book of Changes*, pp. 284–85.

54 A Comparative Study of Chinese and Western Cyclic Myths

Trigram	Image	Attribute	Trait	Direction and season Primal	Direction and season Later	Symbolic Animal	Part of Body	Family Relation
K'un	Earth (Soil)	The Receptive	Yielding Weak Dark	N Winter	SW	Mare (Ox)	Belly	Mother
Chen	Thunder	The Arousing	Active Moving Arousing	NE	E Spring	Dragon	Foot	First Son
Li	The Sun (Fire) (Lightning)	The Clinging	Clinging Depending Beautiful	E Spring	S Summer	Pheasant	Eye	Second Daughter
Tui	Marsh (Heat)	The Joyous	Joyful Satisfied Complacent	SE	W Autumn	Sheep	Mouth	Youngest Daughter
Ch'ien	Heaven	The Creative	Strong Firm Light	S Summer	NW	Dragon (Horse)	Head	Father
Sun	Wind (Wood)	The Penetrating (The Gentle)	Gentle Penetrating Flexible	SW	SE	Bird	Thigh	First Daughter
K'an	The Moon (Water) (Rain)	The Abysmal	Enveloping Dangerous Difficult	W Autumn	N Winter	Pig	Ear	Second Son
Ken	Mountain (Cold)	The Resting (Keeping Still)	Resting Stubborn Unmoving	NW	NE	Dog	Hand	Youngest Son

The next stage in the development of *The Book of Changes*, that of the multiplication of the eight trigrams into sixty-four hexagrams, may be seen as the culmination of the efforts of the exiled Shang diviners to universalize the Pa kua. Certainly there is much historical evidence that the final organization is the work of King Wen (1154–1122 B.C.), founder of the Chou dynasty. It is generally believed that King Wen was also responsible for the arrangement of the Pa kua known as the "Sequence of Later Heaven," a "phenomenal" sequence which presents the Pa kua in the order in which changes are experienced by man in a yearly cycle.

The labours of King Wen (and later, his son, the Duke of Chou) in creating the sixty-four hexagrams provided a sophisticated system of relating the complexities of human behaviour to the processes of change in nature. Arrangement of the hexagrams follows the pattern of natural cycles; thus the penultimate hexagram, "After Completion," represents a gradual transition from ascent to standstill; while the last hexagram, "Before Completion," represents the transition from chaos to order. Its place at the end of the circular sequence points to the fact that every end contains a new beginning.

The Book of Changes presents a complete image of heaven and earth, a microcosm of all possible relationships, for the hexagrams and lines in their movements and changes mysteriously reproduce the movements and changes of the macrocosm. Thus we read that *The Book of Changes* "contains the measure of heaven and earth; therefore it enables us to comprehend the Tao of heaven and earth and its order."[26]

Crucial to any understanding of *The Book of Changes* is the concept of change. In the process of change, every component of the situation can reverse itself according to yin-yang theory and bring a new element into the situation as a whole; moreover, beyond the transformation of opposites, change is also a cycle of phenomenal complexes which are themselves connected, such as day and night, summer and winter, and life and death. The absence of change is still movement, that is, regression, not cessation of movement, for standstill and rest are also aspects of change. Above all change fills the category of time with content, from chaos to cosmos, which is the underlying direction of man's desire to escape temporal limitation.

How does the final formulation of *The Book of Changes* during the Chou dynasty ameliorate the tensions of human ephemerality, the prime motivation behind the systematizing of early Chinese myth and ritual? In the first place change reveals an organic order, for the hexagrams provide complete images of

---

[26] Cited by Hellmut Wilhelm, *Change: Eight Lectures on the I Ching* (New York: Harper & Row Publishers, 1960), p. 69.

conditions and relationships existing in the world. Because human and terrestrial natures obey the same definite laws, and these principles are manifestations of the divine Tao, consistency returns to human life. Secondly, implicit in the concept of change is man's position at the centre of events. Change is neither an intangible snare nor external fate, but an order corresponding to human nature, a guideline from which one can "read" events. The cyclic nature of all movements prevents the movement itself from dispersing beyond human understanding, and becomes, as it were, at the service of man.

Finally, and what is perhaps most crucial, the concept of change defines man's place and responsibility in the cosmos and removes him from subjection to nature. According to *The Book of Changes* man is in a position to intervene in the course of events considerably beyond his own sphere. When, in accordance with the natural order, each thing is in its appropriate place, harmony is established. Now each situation demands the action proper to it, and in every situation there is a right and a wrong course of action; thus the individual comes to share in shaping his fate, for his actions intervene as determining factors in world events. As the centre of events, the individual who is conscious of responsibility is on a par with the cosmic forces of heaven and earth, and in such a manner, can influence changes. In the final analysis we return to Richard Wilhelm (as we once began) to summarize the metaphysical importance of *The Book of Changes*: "It is built on the premise that the cosmos and man... obey the same law; that man is a microcosm and is not separated from the macrocosm by any fixed barriers. The very same laws rule for one as for the other, and from the one a way leads into the other. The psyche and the cosmos are to each other like the inner world and the outer world. Therefore, man participates by nature in all cosmic events, and is inwardly as well as outwardly interwoven with them."[27]

Through its revelation of the universe as organic order and man's position of responsibility as the centre of events, *The Book of Changes* serves as both a summary of and solution to the temporal tensions which were the basis of Chinese myth and ritual. As always, time is of the essence, and the response to the dreadful linearity of human consciousness in the face of natural cycles and divine immortality must be a temporal solution.

Unlike the eschatological solution of the Christians, the Chinese embraced the ever-cyclic natural world as a manifestation of the divine Tao, and through the gradual evolution of a cyclic myth, regained a semblance of the divinity man once enjoyed. The earliest myths and rituals reveal the anguish of human

---

[27] Richard Wilhelm, *The Secret of the Golden Flower: A Chinese Book of Life* (New York: A Helen and Kurt Wolff Book, 1967), p. 11.

ephemerality; and the Myth of the Divine Administration, the Chiao Sacrifices, and the Institution of Ming t'ang suggest the direction of the solution. It remained for the complex abstractions of *The Book of Changes* to provide an ongoing metaphysical system which would satisfy those earliest longings. Of course, *The Book of Changes* could never be the final solution, if there could ever be a "final solution" to the limitations of consciousness. Rather *The Book of Changes* was a plateau of speculation, offering a firm base for the philosophical machinery which soon was brought to bear on the refinement of its central concepts. As the most exact formulation of the cyclic myth yet developed, it became the crossroads of Chinese culture and the source of inspiration for future artists and philosophers.

Before we consider the next stage of the cyclic myth's development at the hands of Chinese philosophers, the following chart is offered to summarize the evolution of ideas discussed thus far. Certainly any attempt to encapsulate three thousand years of Chinese culture on a single page is doomed to ignominious defeat. Moreover this graphic history is perhaps more conceptually accurate than chronologically exact (especially considering the fragmentary nature of the earliest Chinese works). Yet despite these limitations, a visual presentation often succeeds where pages of text cannot, and it makes immediately apparent the importance of *The Book of Changes* as both past summary and future basis.

58 A Comparative Study of Chinese and Western Cyclic Myths

Chronology	Myths	Rituals and Religion				
		God/Ancestor/Heaven	Correspondent Rite	Ming t'ang	Tao/Yin-yang	Pa kua
Mythological Emperors P'an Ku Fu Hsi Nü Wa Shen Nung Huang Ti				Inventor		Inventor
Legendary Kings Yao/Shun/Yü				Tribal thatched hut		
2205–1766 B.C. Hsia dynasty				Astronomical observatory	Tao-mana/moon Lizard clan Polarity concept	Original Pa kua
1766–1123 B.C. Shang dynasty	Creation Universal Destruction Paradise Rebuilt Passages to Heaven Cosmic Rebellion Estrangement/Earth	Shang Ti Worship Ancestor Worship Five-direction Gods	Chiao Sacrifice Feng Shan Other Sacrifices		Correlation of five directions/seasons/ colours/elements	Eight Trigrams Multiplication of Hexagrams
1122–771 B.C. Western Chou dynasty	God of Time Titan K'ua Fu Titan Hou I Goddess Ch'ang O Divine-administration	Non-anthropomorphic Heaven Mandate of Heaven Heavenly Tao/Te		Nine hall cosmic temple	Yin-yang Polarity Five Agents	Eight Trigrams Multiplication of Hexagrams

THE BOOK OF CHANGES
(EIGHT TRIGRAMS—CYCLE OF COSMIC CHANGES)

THE BOOK OF CHANGES

(EIGHT TRIGRAMS— CYCLE OF COSMIC CHANGES, *Chou I*—Early Portion of *Book of Changes*)
(HUNDRED SCHOOLS OF PHILOSOPHY)

	CONFUCIANISM	TAOISM	YIN-YANG SCHOOL	(BUDDHISM)
771–279 B.C. Eastern Chou dynasty				
722–481 B.C. Spring & Autumn Period	Confucius (551–479 B.C.) Tao & yin-yang cosmogony Cosmology—cyclic, organic, harmonious Cyclic time	Lao Tzu (571 B.C.–?) Chuang Tzu (399–295 B.C.) Tao & ying-yang cosmogony Cosmology—cyclic, organic, harmonious Cyclic time	Tsou Yen (305–240 B.C.) Yin-yang & five agent cosmology Cosmology—cyclic, organic, harmonious Cyclic time	Omnipresence of Buddha-nature Nirvana: Attainment of Buddha-nature
403–221 B.C. Warring States Period	Linear time of history & utopian age Correspondence between man and heaven Unity of man/heaven/earth Unity of man and cosmos	Linear time of golden age Correspondence between man and nature Harmony and unity of man and nature	Correspondence between man & universe Unification of man and universe	
221–207 B.C. Ch'in dynasty	THE BURNING OF BOOKS (220 B.C.)			

C. ASPECTS OF CYCLIC MYTH IN CHINESE PHILOSOPHY

1. Philosophy: A Mode of Rationalization

If we accept the view that *The Book of Changes* is truly the crossroads of Chinese culture we must also admit that it is a transition, a crossover from the age of myth and ritual to the era of rationalization and abstraction, and yet a convergence where the four main schools of Chinese philosophy (Confucianism, Taoism, the Yin-yang School, and Chinese Buddhism by the second century A.D.) meet. Throughout history these four schools have been paralleling and reinforcing each other to the point where one can truly say that every Chinese is a Confucian in his social and active life, a Taoist in his individual and passive life, and a Taoist or Buddhist in his religious life, or a Yin-yang adherent in his superstitious life. Fortunately for our investigation, outstanding studies by sinologists such as Fung Yu-lan, Chan Wing-tsit, Nathan Sivin, Joseph Needham, and Amaury de Riencourt are readily available, so we may restrict ourselves to the imprint of the cyclic myth upon Chinese philosophy.

According to Mircea Eliade, "the symbol, the myth, the rite, express, on different phases and through means proper to them, a complex system of coherent affirmations about the ultimate reality of things, a system that can be regarded as constituting a metaphysic."[1] When the mythic mode is exhausted, natural religion is soon superseded by a more discursive and literal form of thought, namely, philosophy; the Yin-yang school of thought occupies this middle ground between natural religion and philosophy.

2. Yin-yang School: Dual-multiple Cosmogony and Penta-cyclical Cosmology

No aspect of Chinese civilization has escaped the imprint of the Yin-yang School's doctrines. Teaching that all things and events are products of Yin and Yang, that phenomena succeed one another in rotation as the Five Elements take their turns, Tsou Yen (305–240 B.C.?) is the central figure of the school and is usually credited with combining the two currents into one. Briefly the concepts of Yin-yang and the Five Elements may be regarded as early Chinese attempts in the direction of constructing a metaphysical cosmology; the former assays the origin of the universe, and the latter the structure of the universe.

As we have already discussed the Yin-yang concept at some length, we should note that the doctrine of the Five Elements is first stressed in the literature of the "Grand Norm" in *The Book of Documents* and the "Monthly Observances" in *The Book of Rites*. Attributed to a speech by the Viscount of Chi of the Shang dynasty towards the end of the twelfth century B.C., the

---

[1] Eliade, *Cosmos and History*, p. 3.

"Grand Norm" provides nine categories through which the Five Elements may be seen to operate. The first category concerns both the substance and nature of each of the elements: Water to moisten and descend; Fire, to flame and ascend; Wood, to straighten and to be crooked; Metal, to yield and to be modified; and soil, to provide for sowing and reaping. When these five in each category come fully and in their regular order, all living things will be rich and luxuriant; if there is extreme excess in any one, disaster will follow. Thus the human and natural worlds were further linked: good or bad conduct on the part of the emperor, for example, would result in the harmony or disturbance of nature.

"The Monthly Observances" is first found in *The Spring and Autumn Annals of Lü* of the third century B.C. It is a small almanac advising what should be done month by month in order to retain harmony with nature. Here the four seasons were correlated with the four compass points and the Five Elements. The interval between summer and autumn became associated with the "centre" as a direction and soil as the element. Again analogies were drawn between human conduct and natural law, and set down as the institution of Ming t'ang, the Hall of Light.

In *A Source Book in Chinese Philosophy* Chan Wing-tsit notes that "by the time of Tsou Yen, the two concepts (the Yin-yang and the Five Elements) were thought of together.... When this interest in correspondence was extended to the realm of political affairs, there emerged a cyclical philosophy of history on the one hand and the mutual influence between man and Nature on the other."[2] With regard to the cyclical concept of history the succession of dynasties accords with the natural succession of the elements. Thus Soil (or Earth), under whose power the Yellow Emperor ruled, was overcome by the Wood of the Hsia dynasty. And the Wood of the Hsia dynasty was overcome by the Metal of the Shang dynasty. The rotation of the dynastic cycle moved from theory to practice in the third century B.C. as the First Emperor of the Ch'in dynasty believed that his "Water" dynasty must govern with harshness and violence to reach accord with the transformations of the Fire Powers. As late as 1911 the official title of the Emperor was still "Emperor Through (the Mandate of) Heaven and in Accordance with the Movements (of the Five Elements)." If such was the influence of Yin-yang teachings on the ruling authorities, one can only imagine its pervasiveness at a more plebian level.

The Yin-yang theory has also put Chinese ethical and social teachings on a cosmological basis. It has helped develop the view that things are related and that reality is a process of transformation. Philosophically it resulted not only in the concept of a common law governing both Man and Nature but also the crucial doctrine of the unity of Man and Nature. To sum up, the Yin-yang School maintained that multiplicity rose from the constant interaction of Yin and

---

[2] Chan Wing-tsit, *A Source Book in Chinese Philosophy*, p 245.

Yang, that the universe is a realm of perpetual cyclic succession of the Five Elements which produce and overcome one another in a fixed sequence, that there is mutual influence between Man and Nature, and that history is a cycle of succession based on the cycle of the Five Elements.

3. Confucianism: The Unity of Man with Cosmos

If one word could characterize the history of Chinese philosophy it would be the "humanism" that professes the unity of man and cosmos. Humanity was a pivotal idea to Confucius (551–479 B.C.) The central concerns of Confucian system were the harmonious universe and the superior man. Humanity, in Chinese "jen," is human nature and righteousness, and the Confucian "Jen" is not a contemplative virtue but an active principle to be carried out; it includes not only all human beings but the universe in its totality. In this connection "Jen" is expanded to the concept of one body with the universe, the generative force of all things and the process of production. The second concept crucial to an understanding of Confucianism is that of Chung Yüng, the Golden Mean. The function of this centrality is to achieve harmony; all differences must be present in their proper proportions. Centrality in the individual is the state of equilibrium in one's mind before feelings are aroused, and harmony is the state after they are aroused. In a societal connotation centrality and harmony together mean complete accord in human relations. Ultimately through the moral principle of Tao, Heaven and Earth and man will attain their proper order; all things will flourish in the harmonious universal operation of the trinity.

The cyclic aspects of Confucianism remained vague and unsettled until the influence of the Yin-yang School. Tung Chung-shu (176–104 B.C.) combined the Confucian doctrines of ethics and history with the ideas of Yin and Yang. For Tung, man was the microcosm and Nature the macrocosm. Nature influenced man because both were governed by the same material forces of Yin and Yang, but man as a replica of Heaven, was superior to all other things of the world. He modified the theory of dynastic succession to a cycle of "Three Reigns": Black, White, and Red, corresponding to the Hsia, Shang, and Chou dynasties. His theory that a ruler rules through the Mandate of Heaven justified the exercise of imperial authority and at the same time set certain limits on it. The Confucianists urged the continual re-examination of imperial rule to ensure that natural harmony was not upset. A fascinating study could be made of the efforts of Tung's Confucianists to lay restraints upon the power of an absolute monarchy.

The scheme of correspondences of the Yin-yang Confucianism of Tung Chung-shu must have held unusual fascination for the medieval Chinese, for it dominated Chinese thought for five centuries. Both Taoists and Confucianists found it congenial because it was a systematized expression of the idea of harmony. However, the real spirit of harmony was lost, whether it was the central harmony of Confucianism, or the inner harmony of Taoism, or the harmony between man and nature as taught by both schools. The doctrine of correspondence soon degenerated into an intellectual sport, a game or puzzle, and finally, a superstition; and was soon replaced by the rationalistic and naturalistic Neo-Taoism.[3]

The next great phase, Neo-Confucianism (960–1912 A.D.), marked a return to the principles of human nature and destiny in *The Book of Changes*. The chief proponent of Neo-Confucianism was Chou Tun-i (1017–1073 A.D.) Elaborating on the cosmogony of *The Book of Changes*, he maintained that in the evolution of the universe from the Great Ultimate through the material forces of Yin and Yang and the Five Elements to the myriad things, the Five Elements were the basis of the differentiation of matter; whereas Yin and Yang constituted their actuality. The two forces were fundamentally one. Consequently the many were ultimately one, and the many had their own correct states of being. The nature and destiny of man and things would be correct in their differentiated state if they all followed the same universal principle. This was the central thesis of Neo-Confucianism for the next several centuries.[4] Neo-Confucianism developed in three different directions: the Rationalistic School of Principle in the Sung period (960–1279); the Idealistic School of Mind in the Ming period (1368–1644); and the Empirical School in the Ch'ing period (1644–1912). We shall restrict our discussion to the cyclic aspects of these schools of thought.

Ch'eng I (1033–1107) and Chu Hsi (1130–1107) were the central figures in the rationalistic movement. Ch'eng I formulated the major concepts and provided the basic arguments, and Chu Hsi systematized Neo-Confucianism into a rationalistic whole. According to Ch'eng I, in the universal production and reproduction the process of daily renewal never ceased. This was a principle to make new things possible. But all principles were basically one, called the Great Ultimate. It was the cooperative functioning of principle and material force that made the universe a cosmos and the fullest realization of "central harmony." Centrality was the order of the universe and harmony was its unalterable law.

In the operation of Yin and Yang, and Heaven and Earth, there was not a single moment of rest in their rise and fall, in their zenith and nadir. The constant succession may suggest that appearance and disappearance followed a cycle, but this cycle should not be understood in the Buddhist sense as a return to the origin. The Neo-Confucian universe was like a vast furnace, and there

---

[3] Chan Wing-tsit, "The Story of Chinese Philosophy," in Chen Charles K. H., compiled, *Neo-Confucianism, Etc.: Essays by Wing-tsit Chan* (New York: Oriental Society, 1969), p. 331.

[4] Chan Wing-tsit, "Chinese Philosophy," in Chen Charles K. H., *Neo-Confucianism, Etc.: Essays by Wing-tsit Chan*, p. 386.

was no such thing as material force returning to its source. Every creation was therefore a new creation, and the universe was perpetually new.[5] When principle was endowed in man it became his nature, which was good, for principle being the utmost source of all goodness. Through moral cultivation selfish desires could be eliminated and the principle of nature realized. The cultivation was by investigation of things, extension of knowledge, sincerity of the will, correctness of feeling, and cultivation of one's personal life. When this was done one would have fully developed one's nature and fulfilled one's destiny. This development of human nature, according to the rationalistic Confucianist, does not stop with personal perfection but involves all things (and here we return to the concept of "Jen"). By interpreting the word in its pun of "the seed or growth," the Neo-Confucianist came to understand "Jen" as the process of production itself, as well as the earlier concept of humanity.

The second or idealistic phase of Neo-Confucianism emphasized only the principle and the investigation of phenomena. The rationalistic Neo-Confucianists regarded mind as a function of man's nature which was identical with principle; but to Lu Hsiang-shan (1139-1193), the central figure of this phase, mind was principle. The mind was originally good and had the innate ability to do good. Filling the whole universe, through all ages and in all directions, there was the same mind. It was identical with all things, for there was nothing outside the way and no way outside things. To investigate phenomena was to investigate the mind; since all principles were complete and inherent in the mind, there was no need to look outside. Lu advocated a simple, easy and direct method of recovering one's originally good nature.[6] A second philosopher of this school, Wang Yang-ming (1472-1529), agreed with Lu Hsiang-shan in the main, but emphasized the direction of the mind, that is, the will. To him, a thing (or affair) was nothing but the mind determined to realize it. Opposed to the rationalistic Confucianist's contention that as things were investigated one's will became sincere, Wang maintained that the sincerity of the will had to precede the investigation of things. For the next one hundred and fifty years Wang's idealistic philosophy dominated Chinese thought, but from the seventeenth century on, Confucianists began to demand the evident, the concrete, and the practical; this marked the beginning of the final empirical phase of Confucianism.

The most outstanding philosopher of the Empirical School, Tai Chen (1723-1777) advocated that principle was nothing but the order of things, and that the way to investigate principle was not through intellectual speculation or even introspection, but by the critical, analytical, and objective study of

---

[5] Chan Wing-tsit, "The Story of Chinese Philosophy," in Chen Charles K. H., *Neo-Confucianism, Etc.: Essays by Wing-tsit Chan*, pp. 342-43.

[6] Chan Wing-tsit, "Chinese Philosophy," in Chen Charles, pp. 392-93.

phenomena based on objective evidence. Despite the "modern" scientific appeal of this concept, Tai was very much a traditionalist in maintaining that the universe was an unceasing process of production and reproduction.

Disregarding the abstract disparities of the various schools of Confucianism, it should be apparent that the philosophy was always a humanism which professed a harmonious universe, and the unity of man and the cosmos. His strong feelings towards humanity led Confucius to expand the idea of "Jen" as the mind of man, the foundation of all goodness, and the source of all production in the universe; and made it possible for the unity of man and the universe to be based on moral principles: the extension of love, the elevation of the mind above the usual distinction between the self and others. By proclaiming man's conduct as a manifestation of the Tao and the determination of universal harmony, he put man inspired by "Jen" in the centre of the cosmos with his integral power to form a trinity with Heaven and Earth.

As a Yin-yang Confucianist, Tung Chung-shu saw that all things had their complements in Yin and Yang which expressed themselves through the medium of the Five Elements, producing and overcoming one another in a fixed sequence. Accordingly he laid the groundwork for the construction of elaborate correspondences of the Five Elements with directions, seasons, colours, tones, and tastes, to underscore the macrocosm. His universe was a harmonious and organic whole; the constant succession of the Yin and Yang was its underlying principle. To him, man was a microcosm and nature the macrocosm; there was a mutual influence between the two. Man was still in the centre of the cosmos and was responsible for the universal harmony, even in Tung's concept of history as a constant cycle of the succession of three reigns.

Emerging from the great infusion of philosophy and Buddhism, Neo-Confucianists differed in their ideas about the means to form the unity with the universe. Some of them, in understanding that the principal law of the existence of all things was One and its manifestations were many, insisted that this unity was based on the understanding of the nature of phenomena through an investigation of things. Others, in understanding that since all principles were inherent and complete in the mind, insisted on a unity of man and cosmos through investigation of the mind. Still others, in seeing that the will was the central function of mind, professed that sincerity of the will had to precede investigation. Nevertheless all Neo-Confucianists were in accord in understanding that the universe was a perpetual transformation, a constant process of unceasing production and reproduction, a spiral of perpetual renewal, and that harmony was the unalterable law of the universe.

As a whole Confucianism reflects the continuation of the cyclic myth in its anthropocentric concept of a harmonious and organic universe; its ever-present theme of eternal return or the periodic regeneration of man and cosmos is strikingly evident.

4. Taoism: The Harmony of Man with Nature

Our third major philosophical school, Taoism, also rose in response to a period of political and social chaos caused by the crumbling feudalism of the Chou dynasty. While Confucianism emphasizes social order and an active life, Taoism emphasizes the individual life and tranquillity. However, it should be noted that the effects of Taoism on Chinese life move far beyond personal quiescence; in its doctrines on government and personal cultivation it is fully the equal of Confucianism.

Originally it was a hermitic school of thought which advocated the preservation of life and avoidance of injury through "escape." Through the work of Lao Tzu (571 B.C.–?) it rapidly evolved metaphysical significance as an attempt to reveal the laws underlying phenomenal change. As these laws remained unchanging, if one regulates one's actions in accordance with them, an understanding of life was assured. A third phase of Taoism concerned the work of Chuang Tzu (399–295 B.C.) and his doctrine of spiritual transcendence as a solution to the problems of mundane life. The Taoism of Lao Tzu begins as one would expect with the Tao (the Way) and Te (the Virtue). Prior to and above all things, "Tao" is the source of all phenomena and the way in which all things pursue their course; "Te" is the potency or virtue obtained from the universal Tao by each individual thing in the process of becoming. "Tao is that by which things come to be, and Te is that by which things are what they are."[7] When Tao is possessed by individual things, it becomes their character or virtue. The individual's ideal life, society's ideal order, and even government's ideal course are based on and guided by the Tao.

The central concept of Lao Tzu's Taoism is "Wu-wei" (Non-action). However, it should not be regarded as negative or quietistic, for Tao is not an escape from the linearity of human consciousness so much as an embrace of the cyclic course of nature. "Non-action" is not inactivity so much as taking no action that is contrary to nature. The Way is a life of simplicity, restricting all activities which are neither necessary nor natural. This idea is grounded in the laws that govern the changes of things, fundamentally the notion that when a thing reaches one extreme, it reverts from it. In accordance with the law of nature, excess is counter-productive. As a way of life then, Taoism denotes simplicity, spontaneity, tranquillity, weakness, and non-action.

The second major figure of Taoism, Chuang Tzu (399–295 B.C.), might be termed the "Philosopher of Changes." To him, Nature is not only spontaneity but a state of constant flux; it is a universal process which binds all things into one, equalizing all phenomena and all opinions. In this life of changes, things

---

[7] Fung Yu-lan, *A Short History of Chinese Philosophy* (New York: The Free Press, 1966), p. 100.

transform themselves by themselves, for this is their nature. A free development of our natures may lead us to a relative kind of happiness, but absolute happiness is achieved through higher understanding of the nature of things. Chuang Tzu's reliance on cyclic process as the Way is best illustrated by his comments on his wife's death: "Now by a further change, she has died. The whole process is like the sequence of the four seasons, spring, summer, autumn, and winter. While she is thus lying in the great mansion of the universe, for me to go about weeping and wailing would be to proclaim myself ignorant of the natural laws. Therefore I stop."[8] Clearly, through his understanding of the priority of cyclic process, the sage is no longer affected by the changes of the world. Thus Chuang Tzu emphasizes the inevitability of natural processes and man's somewhat fatalistic acquiescence to them. Because the Tao embraces all things and unites them, absolute happiness is achieved through complete understanding of the Tao and the identity of man with the universe. In order to do this one must transcend and forget the distinctions between things: life and death; self and non-self; seeing and knowing. As is said in *The Book of Chuang Tzu*: "The universe is the unity of all things. If we attain this unity and identify ourselves with it, then the members of our body are but so much dust and dirt, while life and death, end and beginning, are but as the succession of day and night, which cannot disturb our inner peace."[9] To Chuang Tzu, this is the ideal man, for he not only transcends the ordinary distinctions of things, but also transcends the distinction between the self and the world.

In *The Book of Chuang Tzu*, Taoism reaches a mystical height of transcendence; whereas Lao Tzu's way is directed chiefly to handling human affairs, that of Chuang Tzu aims chiefly at dealing with the universe. The former's goal is social and political reform, that of the latter is transcendence of the mundane world. All in all, Chuang Tzu's contribution to Chinese life was inestimable: his Taoist philosophy helped to transform ancient and medieval Confucianism into Neo-Confucianism, and he was a major influence on the later development of the Zen School of Buddhism. Taoism remained influential until the second century B.C., when its main tenets were codified by Liu An (d. 122 B.C.) in *The Book of Huai-nan Tzu*. Here the Tao is concretized in cosmological terms; Tao originated from a vacuum, and this vacuity produced the universe, which in turn produced the material forces of Yin and Yang. In addition, *The Book of Huai-nan Tzu* synthesized such non-Taoist elements as the Confucianist emphasis on learning as a method of self-cultivation, and the Legalist emphasis on law in government. This approach ensured the survival of Taoism at a time when Confucianism was dominant in government and official thought.

---

[8] Fung Yu-lan, *A Short History of Chinese Philosophy*, p. 108.
[9] Fung Yu-lan, *A Short History of Chinese Philosophy*, pp. 114–15.

The next great resurgence of Taoism was a reaction to the social and political chaos of the waning Han dynasty. Continuous warfare, the degeneration of Confucianism into a scholasticism based on moral and social dogma, and the rise of occultism suggested a spiritual revolt. In the Wei and Chin period, scholars revived the study of ancient philosophers and the first positive result was the notion that "wu" (non-being) should not be regarded as the contrast of being, but as the pure being itself, one and undifferentiated. One facet of this movement, known as the Pure Conversation School, concentrated on a denial of the mundane in favour of romantic wandering, and cultivation of the wit and imagination. The second facet deserves closer attention.

The dominant trend in the Han thought was the correspondence of nature and man and their mutual influence, but the metaphysical schools of the Wei and Chin moved beyond these phenomena to a reality superseding space and time, the non-being of Lao Tzu and Chuang Tzu. These Neo-Taoists were Taoist in their metaphysics, but Confucian in their social and political philosophy. Again, beyond the Confucian correspondences they stressed the overall principle integrating all particular concepts and events.

Despite Taoist contributions to the development of the Zen School of Buddhism in the seventh and eighth centuries, and its influence on the Neo-Confucianists of the eleventh century, the Confucian way of life could not abide the quietism of Taoist philosophy, and Taoism as a philosophical system ceased to exist soon after the first millennium. However, it has profound and lasting influence on every aspect of Chinese culture. It has never lost its domain in Chinese society. Its idea of harmony and simplicity has left deep imprints in Chinese arts, literature, landscape gardening, tea drinking, etc. And as Chan Wing-tsit concludes, "its spirit of naturalism, individualism, and freedom has strongly moulded Chinese life."[10]

It has been stressed how very early in Chinese history the reconciliation of man and time became possible only through identification of linear consciousness with cyclic process. The entire thrust of Taoism was to lend this identification philosophical legitimacy. Lao Tzu discerned an unchanging Tao beneath the unceasing transformations of life. Living in accordance with the Tao, in harmony with nature, and avoiding all excess would result in an understanding of natural processes and thus happiness. Chuang Tzu placed even greater stress on cyclicity, and interpreted the universe as a great current of constant change. By understanding Te as Tao individualized in the nature of things, he professed a transcendence which in its extreme form denied time altogether, and a more spontaneous view of life wherein one was advised to adapt one's own nature to the universal process of transformation and become one with the universe.

---

[10] Chan Wing-tsit, "Taoism," in Chen Charles K. H., *Neo-Confucianism, Etc.: Essays by Wing-tsit Chan*, p. 416.

Finally, although the Neo-Taoists differed in their interpretations of principle as transcendent beyond things, or as immanent in things, and their consequent emphasis on being or non-being; they all agreed that the sage rose above all distinctions and contradictions, and remained in the midst of human affairs and responded to all transformations without discrimination. In this manner the Taoists provided a model of man whose understanding of the Tao beneath cyclic change enabled him to exist freely within the ever-transforming world.

5. Chinese Buddhism: The Identification of Man with the Universal Mind

By the second century A.D. Buddhism, the fourth major philosophical school, had become an influence in Chinese thought. Originally Buddhism was considered a religion of the occult arts, based in part on Hindi tenets. By the third and fourth centuries, as more metaphysical Buddhist texts entered translation, Buddhism was gradually regarded as an equal of philosophical Taoism; and the practice of analogical analysis continued until the fifth century. No doubt the use of Taoist terminology to express Buddhist ideas contributed to a synthesis of the two, and led to the peculiarly Chinese form of Buddhism outlined below.

Although there appear many schools of Buddhism in China, most are united in their treatment of three central concepts—Karma, Samsara, and Nirvana. Because the universe of an individual sentient being is the manifestation of his mind, and every mental event must produce a result no matter how far in the future, some retributive concept is necessary: this is Karma. It is the cause and its retribution is the effect. Thus the being of an individual is composed of a chain of causes and effects. Karma extends beyond one's present lifetime, and this chain of causation is called Samsara, the Wheel of Birth and Death, the main source of man's suffering. A being's essential ignorance of the nature of Samsara leads to worldly attachments despite their illusory nature. Through the teachings of the Buddha, and in the course of many rebirths, the individual may accumulate Karma which does not crave attachment. The result is Nirvana or emancipation from the Wheel of Life and Death. In a word, then, Nirvana is the realization or self-consciousness of the individual's original identification with the Universal Mind.

In its earliest stages in China, Buddhism and Taoism seem indistinguishable. Thus the Pure Land School founded by Hui Yüan (334–416) expresses the hope for rebirth in the Pure Land and is an extension of the Taoist search for everlasting life on earth; the goal was not the termination of human existence, but rather its continuation. A second line of thought, the gospel of universal salvation professed by the great monk Tao Sheng (d.434) maintains that because of the all-pervasiveness of the dharma-body, or the universal existence, of the

Buddha, all things can attain Buddhahood. This doctrine reflects Confucian influence, especially through the older Chinese notion that all people can become sages. At any rate by the time of Tao Sheng Chinese Buddhism developed a theoretical background which led to the formation of three philosophical schools: T'ien-t'ai, Hua-yen, and Ch'an or Zen Schools.

The T'ien-t'ai School, founded by Chih I (538-597) of T'ien-t'ai Mountain, is the first truly Chinese school of Buddhism. It addresses itself to the problems of dharma, or the laws of cosmic or individual existence, and distinguishes a harmony of "Three Levels of Truth." As explained by Chan Wing-tsit, "... all dharmas are empty because they have no nature of their own but depend on causes for their production. This is the Truth of Emptiness. But dharmas are produced and do possess temporary and dependent existence. This is Temporary Truth. Being both empty and temporary is the very nature of dharmas. This is the Truth of the Mean. The three involve each other, for Emptiness renders dharmas really empty, dependent existence makes them relatively real, and the Mean embraces both."[11]

A second characteristic of the T'ien-t'ai School is a rationalist approach to the phenomenal world which finally distinguishes three thousand worlds, the totality of manifested reality. Because each "world" interpenetrates all others, they are immanent in a single thought; that is, all phenomena are manifestations of the Universal Mind and each manifestation is the Mind in its totality. In line with the synthetic nature of much Chinese thought, the T'ien-t'ai School offers a harmony of transcendence and immanence with regard to the Buddha-nature, and leads to the formation of the Hua-yen School of Chinese Buddhism.

The philosophy of the Hua-yen School as expounded by Fa Tsang (643-712) resembles that of the T'ien-t'ai School: each dharma is at once one and all, and the world is in reality a perfect harmony. Consequently when one dharma rises, all dharmas rise with it, and vice versa. In short, the entire universe rises simultaneously. The chief difference between the two schools is the basis for Harmony. In the T'ien-t'ai School the "Ten Characters of Thusness" are invoked through multiplication to produce the three thousand worlds of reality, a process which is not only awkward metaphysically but depends on the mutual inclusion of all dharmas through correspondence and dependence. The Hua-yen School simplifies the unity of dharmas through mutual implication: they imply each other. Since dharmas have no substance of their own, they are empty, and it is this emptiness that combines them through implication. In a real sense, dharmas exist only in relation to each other and to the entire universe, which is a set of interrelationships. Despite the general similarities between the Hua-yen

---

[11] For introduction to the T'ien-t'ai School of Chinese Buddhism, cf. Chan Wing-tsit, *A Source Book in Chinese Philosophy*, p. 396-405.

philosophy and the rising trends of Neo-Confucianism (especially in regard to the one-is-all and all-is-one philosophy), there is one substantial difference. The dominant rationalism of the Neo-Confucianists would not admit the Buddhist notion that all phenomena are manifestations of the mind. Furthermore, through the principle that the universe "produces and reproduces," the Confucians believed that the universe is daily renewed. This creative element is lacking in the Universal Causation of the Hua-yen School.

Turning now to the final flowering of Buddhism in China, the Ch'an or Zen movement has been described by Suzuki as a school in which "the Chinese mind completely asserted itself, in a sense, in opposition to the Indian mind. Zen could not flourish in any other land or among any other people."[12] Although Chinese scholars agree that Bodhidharma (fl. 460–534) did visit China in the early fifth century, Ch'an Buddhism did not establish independent existence until the work of Hung Jen (601–674). By focusing on *The Diamond Sutra*, Hung Jen shifted the emphasis of Chinese Buddhism from the study of Ultimate Reality, or the true nature of dharmas (the cosmic or individual existence), to the human mind itself, or the meditation.

This emphasis on meditation did not so much reflect Indian asceticism as it did Taoist enlightenment through living in accord with the Tao, the way of nature. Two schools of meditation developed: the Northern School, advocating gradual enlightenment through the annihilation of thought; and the Southern School which advocated sudden enlightenment, a notion based on the universality of the Buddha-mind to the extent that any occasion or any moment could serve as a "spring-board" to enlightenment. Historically the Southern School is of greater importance, and many of its ideas were transmitted into Japanese Zen Buddhism. The logical extension of spontaneous enlightenment was to diminish the importance of such typically Indian pursuits as avoidance of worldly involvement, embrasure of intellectual understanding, and the search for unity with the Infinite. In a sense, Chinese meditation with its use of external influence, its worldliness, and its emphasis on wit and insight is a continuation of the original impulse to self-realization through identification with Nature. From the outset, reconciliation with Time depended on involvement with the world of cyclic nature, and in the Southern School of Zen Buddhism this original impulse is recaptured.

Commenting on the humanistic trend of Chinese meditation, Chan has a similar note: "the effect of such strong emphasis on man has been tremendous on Chinese Buddhism. Briefly, it has contributed to the shift in outlook from other-worldliness to this-worldliness, in objective, from individual salvation to

---

[12] Cited by Chan Wing-tsit, *A Source Book in Chinese Philosophy*, p. 425. For introduction to the Hua-yen School of Chinese Buddhism, cf. Chan Wing-tsit, *ibid.*, pp. 406–424.

universal salvation, in philosophy, from extreme doctrines to synthesis, in methods of freedom, from religious discipline and philosophical understanding to pietism and practical insight.... It is also this stress on man that has enabled Buddhism to join with Confucianism and Taoism so that the Chinese can follow all of them at the same time."[13]

In retrospect, Chinese Buddhism differs much from Indian Buddhism in its concept of dharma as emptiness and temporary, for all phenomena are manifestations of the mind of Buddha-nature, and each manifestation is the mind in its totality; thus, as one dharma rises, all dharmas, the universe, rise with it and vice versa. A second difference concerns the idea of harmony, for in the Chinese Buddhism the world of Buddha is a Perfect Harmony and is neither external to the Wheel of Birth and Death nor the phenomenal world, but is here in this present world. And finally, the Chinese concept of Nirvana differs from the somewhat nihilistic approach of the Indian religion, for Nirvana is the gradual or sudden identification with the Buddha-nature which is the true nature of all men and all things. Indeed the Chinese Nirvana is far more in tune with the Taoist unity of man and the universe.

With regard to the cyclic concept of the Buddhist "Samsara," it is tempting to draw several parallels between the Wheel of Life and Death and the eternal recurrence of the cyclic myth. At the most basic level the circle or cycle is an archetype of common occurrence; it is the impulse behind the formation of a cyclic pattern which is crucial to our analysis. The Indian Buddhist impulse is to see man enslaved by Samsara, the Wheel is a function of the illusory nature of the world, and liberation is to escape from its controlling power. In Chinese Buddhism, the essentially negative Indian view is subsumed within a more anthropocentric view which maintains that liberation is possible within the world, even within the Wheel.

6. Summary: Cyclic Myth in the Core of Chinese Culture

From primitive man's first realization of temporal tensions in the myths of Chinese antiquity, to the philosophy of Neo-Confucianism in the eighteenth century A.D., is a sweep of history so vast and deep that any overview must indeed be tentative. Yet certain principles continue to appear, evolve, and appear again; the evolution of the cyclic myth is an elastic concept responding to the temper of the time. Impelled by man's continuing awareness of the disparity between linear consciousness and the cyclic universe, the resolving identification

---

[13] Chan Wing-tsit, "Transformation of Buddhism in China," in Chen Charles K. H., *Neo-Confucianism, Etc.: Essays by Wing-tsit Chan*, p. 434. For introduction to the Zen (Ch'an) School of Chinese Buddhism, cf. Chan Wing-tsit, *A Source Book in Chinese Philosophy*, pp. 425–49.

was first suggested through myth, reinforced and codified through ritual, and finally expanded and lent intellectual legitimacy through philosophy. At no time in this labyrinthine process did the Chinese mind lose sight of the essential anthropocentricity of the cyclic construct; this must be seen as the essential difference between Chinese and Judaeo-Christian culture through the ages. The integrity of the cyclic myth should be evident in the following summary as we review Chinese cosmology, concepts of time and history, and finally the co-existence and correlation of cyclic and linear time in Chinese civilization.

In retrospect, the development of the cyclic concept from myth and ritual to philosophy is mimetic with transitions in the concept of Tao—from a mana, or lunar symbol, to the Way of Nature, and finally to the Absolute Totality, the ultimate assertion of man's oneness with nature and cosmic regeneration. A similar point is made by Nathan Sivin in his "Chinese Conceptions of Time": "Once the Chinese realized that there seemed to be a Way of Nature harmonizing all the pulses of natural phenomena, they came to think of Nature as a great organism itself, with a total life rhythm generated out of the harmony of all its parts, including man."[14] Chinese philosophers, whether Confucianists or Taoists, all seem to agree that ontologically Tao is the infinite substance embracing and uniting all beings, and is also the inevitable destiny of all beings to return to for a peaceful life of virtuous harmony.

Cosmogenetically Tao is the primordial begetter of all things; and the immanent world of beings, in a state of urgent want, will resort to the transcendental world of Tao, in a sense, for the infusion of energy necessary in the performance of adequate function. This endless "Becoming" is the result of the unceasing interplay of Yin and Yang around which all the emblems and symbols are gathered in hierarchial order. The interplay of Yin and Yang evokes and symbolizes the cyclic rhythm of the cosmos and portrays two complementary facets of totality: cooperation and alternation. Amaury de Riencourt compares these processes to the dialectic: "Tao is a synthesis, that which is never quite reached because it always transforms itself into a new thesis which calls for a new antithesis and promotes a further synthesis—and thus on and on, a never-ending process of development."[15]

At any rate, there can be no doubt that Chinese cosmology reflects tremendous influence of a cyclic myth. Whether as the Yin-yang scholar's realm of perpetual cyclic succession of the Five Elements producing and overcoming one another in a fixed sequence; or as the Confucianist's constant process of unceasing production and reproduction, a harmonious spiral of perpetual

---

[14] Nathan Sivin, "Chinese Conceptions of Time," in *The Earlham Review*, I (Fall, 1966), p.86.

[15] Amaury de Riencourt, *The Soul of China* (New York: Coward and McCann Inc., 1958), p. 82.

renewal; or as the Taoist's constant flux of reversal, that spontaneous current which infuses all things according to their natures; or finally as the Chinese Buddhist's perfect one-in-all and all-in-one harmony of the world of Buddha; the Chinese universe always remains a harmonious and organic whole.

The essentially non-mechanistic, non-teleological, and anti-theistic cosmology of the Chinese has informed a humanism which to some Western minds must seem rather naive. Indeed Chinese cosmology not only releases man from mechanistic doctrines of fear and sin because his errors can neither offend personal gods nor threaten his individual existence, but it also offers a most "unthreatening" personal relationship to the cosmos. Evil as an active force cannot exist, nor can it be personified. The implications of such a cosmology tend to make religion superfluous, and religious formalization and institutionalization rather weak, since there is no supreme power knowingly directing the cosmos, and no supreme spiritual entity to be prayed to or implored; but it makes a refined form of magic imperative because man has to be in control of his earthly habitat. By way of summary, Frederick Mote offers a similar conclusion: "The ritualized society of China can be adequately explained in terms of its own cosmology engendering from its cyclic concept, and it is noticeable that this cyclic, harmonious, and organic cosmology has somehow kept man's attention on life here and now and made Chinese thinkers responsible for ordaining the forms and patterns of that life."[16]

It is interesting to note that the word for cosmos in Chinese is "Yü-chou," which has essentially the meaning of space-time. "Yü" denotes all the space in every direction, while "chou" connotes all the time that is yet to come and that has passed since furthest antiquity. The cosmos is thus explained in terms of time and space, or rather as we shall explain further, in terms of man's awareness of his place in time and space.

Attacking a popular Western notion of the Chinese idea of time, Joseph Needham explains: "The culture of China manifested a very sensitive consciousness of time. The Chinese did not live in a timeless dream, fixed in meditation upon the noumenal world. On the contrary, history was for them very real and more vital than any comparably ancient people: and whether they conceived time to contain a perennial fall from ancient perfection, or to pass in cycles of glory and catastrophe, or to testify to a slow but inevitable evolution and progress, time for them brought real and fundamental change."[17]

Chinese temporal thought has always involved two seemingly opposed concepts: there was a cyclical cosmic time without a beginning point, and there

---

[16] Frederick Mote, *Intellectual Foundations of China* (New York: Alfred Knopf, 1971), p. 26.

[17] Joseph Needham, *Time and Eastern Man* (The Henry Myers Lecture, 1964), Royal Anthropological Institute Occasional Paper No. 21 (1965), p. 1.

was a developmental linear time of human history in which man's cumulative cultural achievement had its beginning. The reconciliation of this linear historical time with cyclic time was no doubt made possible by the unique syncretism of Chinese philosophy.

In tracing the development of the institution of Ming t'ang we realize the importance of a calendar which was both astronomically accurate and metaphysically central to the early culture. But according to the nature of Tao, the cosmic calendar (developed through determination of solstices, the lengths of day, month, and year, the motions of sun and moon, and the planetary revolutions) was but one link in a greater, or an infinite, chain of duration. Moving beyond the basic sun-moon cycles the Chinese attempted to build longer cycles, longer rhythms, in efforts to harmonize more and more of the celestial motions. Finally, about two thousand years ago, a total astronomical system was devised, producing a cycle of twenty-three million years, or a World Age.[18] In a sense, they could envisage time beginning anew every twenty-three million years without compromising their conviction that the physical world was eternal. Despite the vastness of this cycle, moments a World Age apart are fundamentally identical, for their Tao is the same; they mark a unique combination of celestial juxtapositions and represent the same beat in the cosmic rhythm.

Within the World Age were subsumed the major theories of dynastic change: the Yin-yang Naturalist's history as a cycle of constant succession in accordance with the transformations of the Five Elements; and the Yin-yang Confucianist's history as a cycle of constant succession of three reigns symbolized by Black, White, and Red. Thus, Chinese history is the manifestation of Tao whose incarnation is a continually renewable process. Yet having professed this generalization, we must deal with the concept of linear time, finely presented by Joseph Needham in his lecture, *Time and Eastern Man*.

Needham discerns three major trends in Chinese thought which support his emphasis on linearity of the temporal view: the historical viewpoint of Confucian humanism, the implicit historicism in concepts of the Golden Age and realizable Utopia, and the rise in Chinese historiography of "continuity history writing."[19] Clearly if a cyclic universe implies the absence of "progress," or lacks definite direction toward better or worse, the Confucian doctrine of the perfectibility of the individual becomes rather unsupportable. Thus the humanity of Confu-

---

[18] According to Nathan Sivin, "When the first complete system of cycles of all five of the planets was worked out, about two thousand years ago, it turned out that the over-all cycle was 23,639,040 years long. It would take that long for the big wheel which drove all of the little wheels to revolve once and a new world age would be starting—when, on a New Year's Day, which is also the first day of the month, the sun, moon, and all the planets were lined up next to each other like a string of pearls." Sivin, "Chinese Conceptions of Time", p. 88.

[19] Cf. Needham, *Time and Eastern Man*, p. 12.

cianism succeeded in establishing a moral code based on an ethical philosophy of history. The temporal thought of a Confucianist would be more concerned with linear limits of human existence. Amaury de Riencourt suggests a similar view in maintaining that "the true goal of the higher type of Confucians was not self-realization through mystical introspection, but the securing of an honoured place in the harmonious procession of historical personages."[20] This notion of "right-timeliness" underlies the popular Western idea of the Confucianist as an ancestor-worshipper.

Early Chinese attitudes toward the development of human society often assumed contrasting forms. The more primitive notion was of a Golden Age of Communalism of Sage-Kings from which mankind had steadily declined. On the other hand we find the idea that these culture-heroes were progenitors of something much greater than themselves, that eventually a Utopian Age of the Great Togetherness and Harmony (Ta T'ung), or the Great Peace (T'ai P'ing) would develop. Both views are united in their implied opposition to feudal society, and their concern with historical change, albeit within the vast matrix of the World Age cycle.

A third essentially linear development in Chinese temporal thought is distinguished by Joseph Needham in the rise of Chinese historiography.[21] As Confucian historians progressed beyond the usual concerns of dynastic legitimacy, there evolved various forms of "continuity history writing" which dealt with long periods of time involving several dynasties, and this manner overcame the compartmentalization of time advocated by earlier Yin-yang scholars.

Of course these linear aspects of Chinese temporal thought were always affected by (if not subsumed within) a more cyclic viewpoint. Indeed Joseph Needham's point is not the primacy of the linear perspective, but the development of historicism in China, a development seen necessary by some to make the leap between civilization and culture. Whatever the concerns of Confucian historians regarding imperial rule, the rulers themselves were governed by a cyclical world-view embodied in the institution of Ming t'ang. Perhaps the most reasonable approach would be to minimize the dispute over where Chinese culture stood in the contrast between a linear perspective of time and the myth of eternal recurrence by admitting that both kinds of time were and are coexistent, or at least important on different levels: the former relevant in moral, social and historical contexts, while the latter in spiritual, artistic, philosophical, and metaphysical perspective. Naturally, it is the literary level which will be the focus of the next chapter.

---

[20] Riencourt, *The Soul of China*, p. 15.
[21] Cf. Needham, *Time and Eastern Man*, pp. 11-12.

A quotation from a contemporary Chinese scholar, Thomé H. Fang, may serve by way of a summary of the Chinese conception of time:

> "The question is, What is time? The essence of time consists in change; the order of time proceeds with concatenation; the efficacy of time abides by durance. The rhythmic process of epochal change is wheeling round into infinitude and perpetually dovetailing the old and the new so as to issue into interpenetration which is continuant duration in creative advance. This is the way in which time generates itself by its systematic entry into a pervasive unity which constitutes the rational order of creativity. The dynamic sequence of time, ridding itself of the perished past and coming by the new into present existence, really gains something over a loss. So, the change in time is but a step to approaching eternity, which is perennial durance, whereby, before the bygone is ended, the forefront of the succeeding has come into presence. And therefore, there is here a linkage of being projecting itself into the prospect of eternity."[22]

The evolution of the cyclic myth through ritual and philosophy must be seen as a triumph over time, as a victory over the earliest temporal tensions of primitive man. If myth first adumbrated these tensions, if ritual was to ensure that the rhythm of man's life on Earth was in full accordance with the rhythm of Heaven, if philosophy lent intellectual legitimacy to the unity of man and the cosmos; then this was the Tao, complete and fully understood. The motivation has always been to triumph over time. "This is why Confucians have craved so much for the continually creative potency of the heavenly Tao in the shaping of the cosmic order as a whole. This is why the Taoists have whole-heartily cherished the ideal of nothingness for its coming to the rescue of all things relative in the realm of Being. And this is also why Chinese Buddhists have vehemently struggled for the partaking of the Buddha-nature embedded in the integral truth of the ultimate spiritual Enlightenment."[23] The three major philosophies of China shaped the cyclic myth into a cultural symbol whereby man's place in time and space was readily discernible. The unity between man and nature, as represented by this cultural symbolism, reconciled primitive temporal tensions with more advanced experiences and discoveries.

---

[22] Thomé Fang, "The World and the Individual in Chinese Metaphysics," in Charles Moore, ed., *The Chinese Mind* (Honolulu: University of Hawaii Press, 1967), p. 240.

[23] Fang Thomé H., "The World and the Individual in Chinese Metaphysics", p. 259.

D. ASPECTS OF CYCLIC MYTH IN CHINESE LITERATURE

1. The Mythic and Philosophical Aspects of the Poems of "Encountering Sorrow" and "The Return"

Myth, as a mode of cognition or a narrative resurrection of unconscious drives, wishes, fears and conflicts, circumvents a verbal universe in which man's psychological, philosophical, religious, social, and even political histories are really one in the juxtaposition and succession of the stages of gods, heroes, and men. As rites degenerate, myth merges inseparably with literature, art, religion, and various other symbolic forms. Thus Northrop Frye notes that "there are two structures in a culture which descend from mythology: one is literature, which inherits the fictional and metaphorical patterns of mythology, and the other is a body of integrating or cohering ideas, also mainly fictional, in religion, philosophy, and kindred disciplines."[1] In our study of Chinese literature it is truly impossible to separate the cyclic myth and its developments from the works studied. Indeed, by definition, within a homogeneous culture such a separation would be unthinkable.

The history of Chinese literature properly begins in the fifth and fourth century B.C. in the regions of the Yellow River and the Yangtze River, the twin cradles of ancient Chinese civilization. Concerning the poetry of the former area, it is said that Confucius (551–479 B.C.) selected some three hundred poems into an anthology, *The Book of Odes*. Composed mainly of ballads and festal songs, *The Book of Odes* rightly belongs in the company of the Veda, Homeric epics, and Psalms; perhaps more so with the latter because of its intimate expression of the voice and feelings of the common people. It maintains a spontaneous confidence in life which later Chinese poetry never really recaptures, and has since become the core of Confucian literature.

Whereas Chinese history had its beginning in the northern region where we first find reference to sage-kings and dynastic rulers of antiquity, Chinese mythology found a favourable climate for development in the southern or the Yangtze area. Unlike the harsher northern surroundings, the fertile Yangtze region enabled its residents to live in comparative ease, to indulge in dreams of the romantic and the supernatural. These factors gave rise to a literature of metric songs different from *The Book of Odes* in their lyric nature and romantic spirit. Indeed *The Songs of the South* is more sentimental, even self-conscious, and often evokes the supernatural and the other-worldly. Together these two texts constitute the original mainstream of Chinese poetry.

The greatest solo voice of *The Songs of the South* was Ch'ü Yüan (343–277 B.C.) His "Encountering Sorrow" (Li sao) is the earliest long narrative poem to survive. A brief biography reveals a life of political upheaval and exile which

---

[1] Frye, *A Study of English Romanticism*, p. 5.

rivals that of Dante. As a member of the ruling house of the state of Ch'u, his brilliant diplomatic career was shadowed by the jealous intrigues of his fellow ministers. His first banishment in 305 B.C. occasioned many of his most brilliant lyric songs: "Encountering Sorrow," "Outpouring of Sorrow," and "Inquiry into the Cosmos." A second exile in 286 B.C. resulted in his "Summoning the Soul," "Thinking of the Fair One," and "Crossing the River." The capital of Ch'u was finally plundered and ruined by the conquering army of Ch'in in 278 B.C., and at the age of sixty-six Ch'ü Yüan composed "Lament for the Capital," and "Embracing the Sand," expressions of his rejection of patriotic hope. According to legend, one year later, on the fifth day of the fifth moon he committed suicide by drowning. As the first great poet of China, Ch'ü Yüan has been called the father of Chinese poetry and has become a cultural hero. In some southern territories he has even been worshipped as a water-deity. At any rate, the day of his death is designated as Poet's Day, and the Dragon Boat Festival commemorates his drowning.

a. "Encountering Sorrow" (Li sao): A Mythic Quest for the Unity with Cosmos

Generally speaking, "Encountering Sorrow" is allegoric and often anagogic. It consists of two parts: the tristesse and the quest, which David Hawkes designates as the "tristia" and the "itineraria."[2] The tristia expresses the poet's sorrow, complaint, and resentment against a malicious society which through slanderous misrepresentation has separated him from the Fair One he serves. The itineraria describes the poet's journey in search of a new mate, perhaps a goddess or legendary beauty, and finally the journey's end whereby spiritual transcendence triumphs over embittered despair and anguished nostalgia, as outlined below:

"Encountering Sorrow" begins with the poet's declaration of divine ancestry through Kao Yang, or Chuan Hsü, the grandson of the Supreme Being and the Sovereign God of the North, who had supervised the Estrangement of Earth from Heaven. This declaration justifies the derivation of the poet's oracular names, "True Exemplar," and "Divine Balance."(1-2)[3] In the second section the poet catalogs his innate and cultivated virtues, then claims that he is in "fearful pursuit" of Time. The problem appears to be one of reconciling the diurnal and seasonal cycles with which the poet clearly identifies, to the longer cycle of a human lifetime, specifically the ephemerality of youth and love.

---

[2] Cf. David Hawkes, "The Quest of Goddess," in *Asia Major*, XIII: 1, 2 (1967), p. 82.

[3] Cf. David Hawkes, trans., *Ch'u Tz'u: The Songs of the South* (Boston: Beacon Press, 1962), pp. 21-34. The numbers in the bracket refer to the quatrains found in his translation.

Using the narrative device of direct quotation, the poet next recalls his encouragement of the Fair One to follow him (6-12). He cites examples of legendary kings who also followed a virtuous path and so preserved their peace and beauty. His speech is a failure; the Fair One remains inconstant, and the poet is exiled. Cynically he proclaims how "loyalty brings disaster," yet invokes the Ninefold Heaven to witness his enduring loyalty.

In resignation and as consolation, he devotes himself to the cultivation of flowers: his garden is as large as it is varied (13-14). At this point the blossom imagery begins to serve many purposes, for the cultivation of flowers is linked to the personal cultivation necessary to achieve liberation from sorrow. Moreover there is a constancy within the vegetative cycle which the poet contrasts with the inconstancy of the mundane world. Through identification with cyclic nature, the poet believes his "mind can be truly beautiful." By clothing himself with fragrant flowers, he follows the solemn way of P'eng Hsien, an ancient sage recluse, and attempts to transcend the ephemeral. Finally, the poet offers an anguished contrast between his way of life and that of a "generation of cunning artificers" (22-26). He will bear blame and endure insults, but also keep himself pure and spotless and die in righteousness. There follows a moment of vacillation, but again the poet resolves to continue on his virtuous path: he will traverse the four quarters of the world to gather flowers and fragrances to fulfil his constant love for beauty and nourish his immanent virtue. Here Ch'ü Yüan makes explicit the Taoist notion that love of nature leads to transcendence of the material world: "Even if my body were dismembered:/ ...how could dismemberment ever hurt my mind?" (27-32).

An encounter with Nü Hsü, Goddess of Matchmaking,[4] fails to alter the poet's course as she warns that the consequence of persistence is often death, and attempts to bring about a compromise between the poet and the sophisticated world (33-35). In reply, he appeals to the ancient sage-king, Shun, for inner guidance, and following an inventory of the rise and fall of dynasties, concludes that the high god in Heaven knows no partiality and allows only the good and virtuous to flourish. The poet therefore decides to endure his isolation from the mundane world and his estrangement from the Fair One (36-44). This confirmation signals the end of the tristia; in a mood of embittered despair the poet completes his anxious preparations for the celestial journey to come.

The itineraria begins with the poet's celestial journey to the cosmic mountain, K'un Lun.[5] In Chinese mythology, Mount K'un Lun is the abode of perfect blessedness and the passageway to Heaven. It consists of five circular levels along a spiral path but is considered beyond mortal reach. Nevertheless

---

[4] Pi Hai-chi, *Wen-hsüen yen-chiu hsü-chi* (On Literature) (Taipei: Shang-wu yin-shu-kuan, 1971), pp. 55-56.

[5] Cf. II-A-1-f, for the Myth of Mount K'un Lun.

the wandering poet quickly ascends to the Hanging Gardens, the uppermost circle—another indication of his innate beauty and cultivated virtue. It is a spiritual realm of immortality, and by ascending it one may command the wind, the rain, and the lesser gods and goddesses. Perhaps reminded of mundane ephemerality, the wanderer's first command is for Time to stop. He waters his dragon-steeds at the Lake of Purity, and rests under the cosmic Fu Sang Tree,[6] before rushing his glittering train to the gates of Heaven, the realm of pure divinity. Here his progress is checked by the Superintendent of the Nine Heavens, Lu Wu.

Through a night of indecision, the poet mends his orchid garments and crosses the White Water, one of the colourful rivers flowing out of the Lake of Purity where Hsi Wang Mu, the Supernal Mother Goddess, dwells. He ascends Mount Cool Wind in the fourth circle, the realm of immortality, and despairs again for there is no fair lady (53-54). Sending one of his attendant gods in search of a nymph, he remains in the House of the Spring where Ch'ung, the assistant to the Sovereign God of the East and the son of the Sovereign God of the West, is the God of Spring and Life.[7] Despite the efforts of the Patroness of Marriage as a go-between, the nymph proves to be a rather vain hedonist and very difficult to woo. The poet resolves to seek elsewhere for his mate (55-58).

Our wanderer scours the heavens and the four quarters of the earth before finding the jade tower in which Chien Ti, a legendary beauty and later the ancestress of the House of Shang, is confined by her father, the Lord of Sung. As homage, the poet sends a magpie, then a phoenix, yet fails in competition with his virtuous rival, the first ancestor of the House of Shang (59-61). At this point the poet turns to woo the Lord of Yü's two princesses, but they live in virtue and are to marry Prince Sao K'ang. The wanderer laments the inaccessibility of sage-kings and their daughters the world over, and despairs his present situation (62-64). For consolation, the poet seeks out sacred stalks and approaches Ling Fen for a divination. An auspicious oracle is divined: "Beauty is always bound to find its mate./ Who that was truly fair was ever without lovers?" (65-69). Following the diviner's encouragement to seek elsewhere for a truly fair lady, the poet takes this opportunity to renew criticism of the follies of the mundane world. In indecision he consults the spirit of Wu Hsien, the chief of the shaman ancestors.[8] After enumerating historical examples, Wu Hsien addresses the poet in oracular style:

---

[6] Cf. II-A-1-j, for the Myth of Fu Sang and Ten Suns. Cf. II-A-1-f, for the Myth of the Lake of Purity.

[7] Cf. II-A-1-f, for the Myth of Mount Cool Wind. Cf. II-A-1-n, for the Myth of Hsi Wang Mu. Cf. II-A-1-o, for the Myths of the God of East and Spring.

[8] Tai Chen, "Ch'ü Yüan fu chu" in *Ch'ü-tz'u ssu chung* (Commentary on Ch'ü Yüan's Poetry) (Hong Kong: Kuang chih shu-chü, 1959), I, 8.

> "To and fro in the earth you must everywhere wander,
> Seeking for one whose thoughts are of your own measure...
> As long as your soul within is beautiful,
> What need have you of a matchmaker?...
> Gather the flower of youth before it is too late,
> While the fair season is still not yet over." (70-75)

In retrospect the poet comes to a bitter realization: that the world is a disordered tumult of changing, and drifts to conform to evil counsel; that all fragrant flowers such as orchids and peppers have been transformed into worthless mugwort; all because they have no persistent care for beauty. With only his garland of pristine fragrance left, the poet decides to follow Ling Fen's advice and transcend the moribund world by continuing his journey in quest of a mate (76-85).

In his jade and ivory chariot, accompanied by an increased retinue, the poet departs for Mount K'un Lun and the world's western extremity. In Chinese mythology, beyond the Dark Water lies the Crescent of Sinking Sand, the western boundary of Earth and Heaven. Beyond lie the sacred mountains K'un Lun, Ch'ang Liu, and Yü, the abodes of the Gods of Completion and Perfection, the place of sunset and sundown. It is the boundary toward which Titan P'eng Tsu at the age of eight hundred travels in search of immortality.[9] Crossing the blood-red Sinking Sand on a bridge of dragons, the poet soars through the Nine Underworlds and the Nine Heavens in the poem's most floridly descriptive passage. But a sudden glimpse of his old home in the distance causes both groom and dragons to refuse to go on. The journey is brought to an abrupt end.

There remains only an Envoi (93), in which the poet reaffirms his decision to remain in exile, and vows to continue after P'eng Hsien's example: to quest for sacred plants and divine blossoms, and finally, a soul-mate. The poem ends with a triumphant restatement of the value of virtuous cultivation and reclusiveness as the path to transcendence.

Before elucidating the cyclic aspects of "Encountering Sorrow," it is necessary to provide some background to the intellectual milieu in which it was written. The Third Century B.C. in China was a time of disillusionment and self-realization, a time when magic, ritual, and myth had not been superseded by rationalism and abstraction. Myths of ancient gods and goddesses existed simultaneously with the rising systematization of the Yin-yang school, or the abstracting tendencies of Taoism and Confucianism alike. Certainly, ancestor worship and the use of Pa Kua, the precursor to *The Book of Changes*, in divination and philosophy were still popular. In fact the use of magic and rite

---

[9] Cf. II-A-1-k, for the Myth of the Sinking Sand and Titan P'eng Tsu's Sorrow.

still prevailed in the southern regions, especially among the barbaric peoples where Ch'ü Yüan was exiled. With this in mind, it is no surprise that the identity of the poet as a mystic, a Yin-yang disciple, a Confucianist, or a Taoist has been disputed for centuries. It is only recently that scholars have reached consentient agreement on the profusion and profundity of Ch'ü Yüan's sources.[10] Thus the traditional bureaucratic interpretation of "Encountering Sorrow" has yielded to the study of **sensus spiritualise**. "Encountering Sorrow" holds a mirror to its turbulent age, and is a condensed song of the various speculations of the era.

Structurally speaking, the tristia of "Encountering Sorrow" consists of a realization and a decision which require the search for fragrant blossoms and the quest for a mate, which in turn constitute the progress of the itineraria. This realization develops from the tragic sense of the poet's self-consciousness upon his separation from the Fair Lady, his solitude in the mundane world, and his estrangement from the cosmos. The notion that the poet is isolated in time as well as space is reinforced by images of the "swift steeds" of Time and the ephemerality of life, manifested in the periodic changes of seasons and the incessant transformations of the vegetable world. This crucial awareness creates an embittered nostalgia and an anguished desire for reconciliation; both demand self-cultivation and self-fulfilment through vegetable adornment as a means of obtaining a unity with the Tao, fulfilling the quest for a mate, and ensuring the progress of the cosmic journey.

The Chinese have always believed that there is a mutual relationship between man and cosmos, and it is in this connection that evergreen stalks, beautiful flowers, and fragrant blossoms are understood to contain a sacred quality capable of bestowing purity, perfection, and even divinity. The symbolism of vegetable adornment in "Encountering Sorrow" has a shamanic origin from *Chiu ko* (The Nine Songs), the erotic liturgy of the southern barbarian which was refined and edited by Ch'ü Yüan under royal command.[11] The vegetable world in "Encountering Sorrow," as in the liturgy, is one of sharp contrast between such rank weeds and pale flowers as mugwort and dogwood, which symbolize the slanderous and the profane, and such beautiful and fragrant flowers as orchid and pepper, which symbolize the virtuous and the divine. In formative trope, the latter flowers become a configuration of the poet's innate virtue as a descendent of the Sovereign God of the North.

---

10 Cf. Wu T'ien-jen, *Ch'u-tz'u wen-hsüeh t'e-chih* (Literary Characteristics of the Songs of Ch'u) (Taipei: Shang-wu yin-shu-kuan, 1972), pp. 12–45. Also cf. Chan An-t'ai, *Ch'ü Yüan* (Shanghai: Jen-min ch'u-pan-she, 1957), pp. 60–72.
11 Cf. Su Hsüeh-lin, *Chiu ko chung jen-shen lian-ai wen-ti* (The Problem of Love Between Mortal and Immortal in The Nine Songs) (Taipei: Wen-hsing ch'u-pan-she, 1967), pp. 35–37.

Moreover, once the poet confronts his self-conscious awareness, they become an emblem and a pledge of his persistent self-cultivation of virtue and incessant pursuit of unity with the "fair lady" and the cosmos. Finally, upon the defeat of the orchid in its allegorical battle with the weed, the emblem becomes the primary motif of the love quest and cosmic journey.

With regard to the itineraria we find two major themes, the quest itself and the consequent "progress" within the quest. Both concepts share resemblances to the Myth of Mount K'un Lun, the Feng Shan Ritual, and the practices of the Ming T'ang Institute. At the heart of the poem is the resolution of temporal tensions (symbolic in exile) through reconciliation and union with the divine. Whether we speak of the emperor's mimetic cosmic journey through monthly, directional, and seasonal progression in the symmetrical microcosmic Ming T'ang; or his sacred journey and thanksgiving sacrifice to Mount K'un Lun or other cosmic mountains; or solar and lunar symbols of regeneration in the sacrifices to Heaven, Earth and the Four Directions; each is concerned with re-establishing a unity with the cosmos, re-harmonizing the powers of Heaven, Earth and Man, and partaking of the quest and progress as a cosmological reconciliation"

As David Hawkes stated, the idea of the progress is magical, for a complete and successful circuit of the cosmos will make one a lord of the universe, able to command any of its powers at will.[12] Thus the poet, as the representative seeking a shaman or emperor, undertakes a ritual journey consisting of an ascent to the hub of the cosmos, then a circuit of the various quarters of the mandala-like universe, and finally a return to the centre of power. The wanderer is initially successful, but the refusal of Heaven's guardian to admit the poet (47-52) indicates that reconciliation is not yet possible, that the circuit must be completed within the quest. Again, the symbolism of beauty and the feminine in "Encountering Sorrow" has much to do with primitive thought. The metaphoric kenning of flower to virtuous king or beautiful goddess is found in *The Book of Odes* and the liturgical ritual of *The Nine Songs*; while the moon-earth-woman configuration as the regenerative feminine element requiring the masculine complement for virtuous unity is evident in some shamanic rituals and especially in *The Book of Changes*[13]. Despite scholarly dispute as to the specific allegory of the goddess or beauty (whether as the goddess in myth, the nymph in shamanic liturgy, or the virtuous king or way of humanity in Confucian interpretation,) a common ground is indicated by returning to the original cyclic myth with its impulse toward temporal reconciliation, and its emphasis on the periodic alternation of opposites.

---

[12] Cf. Hawkes, "The Quest of the Goddess" p. 82.

[13] Again a primitive universality is supported by similarities to such Jungian archetypes as the heroic quest for the "anima."

The love quest is undertaken in the correct circuit according to the progress of due direction. Although the quest is not fulfilled, due to the necessity of the allegory resembling the poet's futile mundane pursuit, the circuitous progress has assured him a harmony with the cosmos and granted him re-ascendance to the hub of the universe. Spiritual success is indicated by his transcendental flight between the Nine Earths and the Nine Heavens, and in the reception of the Nine Heavenly Hymns and the Nine Divine Dances. While the poet's nostalgia for his ancestral home in the capital of Ch'u underscores the inevitably unsuccessful quest in the love allegory, and brings the tragic futility of his piety to a climax, it in turn reinforces the triumph of spiritual transcendence and the way of the sage recluse.[14] Metaphorically speaking, the poet is never closer to "home" than when he is soaring at the western extreme of the universe in harmonious unity with the cosmos.

In a word, the poetic **amour** and **pietas** of the romance find their allegorical expression and formative tropes in the mythos of the quest and progress; each in turn may be seen as abstractions of the cyclic myth, the paradoxical centre of the poem. Exiled from all that he loves, the poet focuses on the decadent aspect of cycle with regard to society, and the incessant aspect of cycle with regard to the time of natural succession. This, in turn, leads him to lament the linear aspect of human life. Paradoxically it is the obverse cyclic aspects which provide his liberation: the re-creative side of natural cycle evident in his concern for vegetable garments; and his use of incessant change to provide both the model for his continued wanderings, and the constancy necessary to a transcendent understanding of human life in the Taoist sense; and finally his reconciliation with time and the cosmos.

In the formation of a literary genre, "Encountering Sorrow" has influenced other poets appearing in *The Songs of the South*, and poets of succeeding generations in both style and spirit. Prosodically, its emblematic density contributed much to the development of the "Rhapsody" in the literature of the Han dynasty. Thematically speaking, the cosmic progress and the quest of the goddess reappear in such poetic works as "The Rhapsody of the Goddess" by Sung Yü of the Warring States period, "The Celestial Journey of the Great Man" by Ssu-ma Hsiang-ju of the Western Han period, "The Rhapsody of the Nymph of River Lo" by Ts'ao Chih of the Three Kingdoms period, and in poems of mystic journeys by Kuo P'u and his contemporaries in the Chin dynasty. David Hawkes has noted in his study of the quest archetype in Asian literature that the panoramic enumeration of cosmic progress in "Encountering Sorrow" and the liturgy of shamanism has in due time become a cosmological

---

[14] Further discussion of the romance elements in "Encountering Sorrow" may be found in Yeh Shan, "Yi-shih yü chuei-ch'iu" (Emblem and Quest), in *Ch'un wen-hsüeh*, 3:6, pp. 52–53.

approach to art and literature. Thus in the third century A.D., Lu Chi's "Rhapsody of Literature" describes the creative writer as a poet-magician, an itinerant mystic who explores the universe to acquire the powers of literary creation:[15]

> "Taking his position at the hub of things, the writer contemplates the mystery of the universe....
> His spirit gallops to the eight ends of the universe; his mind wanders along vast distance.
> In the end, as his mood dawns clearer and clearer, objects,
> Clean-cut now in outline, shove one another forward.
> He sips the essence of letters; he rinses his mouth with the extract of the six arts.
> Floating on a heavenly lake, he swims along; plunging into the nether spring he immerses himself."[16]

With regard to the tristia, although it has no similar generic development, through the poetry of Sung Yü it has received the focus of "sorrow over autumn." Thus in Chinese literature the image of autumn has become not only an allusion to the tristia of "Encountering Sorrow," but a symbol of man's temporal awareness of the ephemerality of being, of his realization of his estrangement from the cycle of eternal return.[17]

b. "The Return" (Kuei ch'ü lai tz'u): A Pastoral Quest for Harmony with Nature

Similar to the Warring States period background of "Encountering Sorrow," political turmoil, warfare, and natural disasters of the third and fourth centuries A.D. form the background to the next poet under study, T'ao Ch'ien (365–427). In desperation the Chinese turned from the romantic poetry of itinerancy and the steadily retreating mythological worldview to embrace that syncretist movement which saw the rise of Chinese Buddhism and the synthesis of Confucianism and Taoism with regard to the unity of man and nature. The exemplary man became the sage in intimate companionship with nature, who rose above all distinctions

---

[15] For further information on Sung Yü, Ssu-ma Hsiang-ju, Ts'ao Chih, cf. Lin Wen-keng, *History of Chinese Literature* (Taipei: Kuang-wen shu-chü, 1963), pp. 63, 76, 87–89; on Kuo P'u, cf. K'ang P'ing, "Lun Wei-Chin yu-hsien-shih te hsing-suai yü lei-pieh" (Study on the Rising and Falling, and the Classification of the Poetry of the Quest of the Gods in Wei Chi period) in *Chung-wai wen-hsüeh*, 3:5, pp. 154–58.

[16] Achilles Fang, "Rhymeprose on Literature: The Wen-fu of Lu Chi," in *Harvard Journal of Asiatic Studies*, 14 (1951), pp. 531–32.

[17] Cf. Lin Wen-keng, *History of Chinese Literature*, pp. 63–87; Hawkes, *Ch'u Tz'u: The Songs of the South*, pp. 92–100.

and contradictions, but remained in the midst of human affairs and responded to all transformations spontaneously without discrimination. In literature, the spiritual reclusiveness of "Encountering Sorrow" coupled with the diminution of confidence in life inherited from *The Book of Odes* engendered a realm of tranquillity accessible to earth-bound mortals. A major preoccupation was the attainment of peace of mind which would enable one to enjoy contentment within one's meagre existence.

T'ao Ch'ien lived through the turbulent waning of the Chin dynasty, a half-century of revolution, banditry, and regicide. One of China's truly great writers, he initially attempted to use his education in service to the state but was unable to compromise his principles to the corrupt bureaucracy of the time. In 405 A.D., he resigned his magisterial post and embraced the life of the peasant. This was the occasion of his rhapsody of "The Return."[18] Crop failures and fires could not convince him to exchange relative poverty for the barren rewards of respectability, and his poetry survives as a testament to the joys of living in simple harmony with nature. T'ao Ch'ien's poetry best expresses the dilemma of a man of good will born into the troubled times of medieval China. As a poet and recluse he does more than give meaning to a particularly chaotic period of Chinese history; he belongs to that small group of poets who are properly called philosophical, who crystallize attitudes toward life that are valid in other times and places. He has long been recognized by Chinese critics as the master of recluse poetry and the father of Chinese pastoral poetry. His unique philosophy of natural harmony and assimilation with cosmic change is best illustrated in his rhapsody of "The Return."[19]

On one level "The Return" is a revelation of a gradual fulfilment of man's spiritual quest for eternal return. Perhaps, recalling "Encountering Sorrow", it exhibits a binary structure; the call to return and the return itself. Through severe self-examination within a **"tristia"** similar to that of "Encountering Sorrow," the poet discovers his situation to be that of a gardener overcome by the weeds of existence, an allusion to the waste garden of orchids in "Encountering Sorrow"; this is the immediate motive behind the poet's decisive urge to return.

---

[18] "Kuei ch'ü lai tz'u," the Chinese title of the rhapsody means "to go away and to return home." Lucian Miller translated it into "Homing" in his remarkable article, "Poetry as Contemplation: T'ao Ch'ien's 'Homing' and William Wordsworth's 'Tintern Abby'," in *The Journal of the Institute of Chinese Studies*, 6:2 (University of Hong Kong Press, 1973), pp. 565-84. James Robert Hightower translated it as "The Return" in his outstanding work, *The Poetry of T'ao Ch'ien* (Oxford: Clarendon Press, 1970), pp. 268-70. Couplets or quatrains cited hereafter are adopted from Hightower's translation.

[19] For the biography of T'ao Ch'ien, cf. Hightower, *The Poetry of T'ao Ch'ien*. pp. 1-6; also cf. Chang Chih, *T'ao Yüan-ming chuan-lun* (Shanghai: Hsüeh-ti ch'u-pan-she, 1953), pp. 111-24.

The ruined garden for which the poet grieves may be seen as the corrupted mundane world whose attachments have bound the poet to this point in his life. His bitter disillusionment has developed through a lifetime of frustrations, over the historical truths revealed by ancient sages and virtuous kings, and even over the Confucian doctrine of service and moral duty to mankind. The sense of homelessness which the poet feels in bureaucratic service is best expressed in another of T'ao's works, the fourth of "Twenty Poems After Drinking Wine," through the image of the lost bird.

> "Anxious and seeking, the bird lost from the flock-
> The sun declines, and still he flies alone,
> Back and forth without a place to rest;
> From night to night, his cry becomes more sad,
> A piercing sound of yearning for the dawn,
> So far from home, with nothing for support... ."[20]

The plaintive sound of the bird yearning for its nest is echoed in the poet's profound nostalgia for home. To the poet the homecoming call demands reclusion from the sophisticated world:

> "By mischance I fell into the dusty net
> And was thirteen years away from home.
> The migrant bird longs for its native grove.
> The fish in the pond recalls its former depths."[21]

What is most important is the direction of "home"; return does not point to Mount K'un Lun of mythology, nor to the archaic Golden Age, but to the present, to the family home in the forest of the southern hill. The return is neither divine nor mythic, but human and existential. The minutiae of pastoral existence circumscribe the destiny and the immanent content of "the Return to Nature." Nature, as the poet and his contemporary syncretic metaphysicians comprehend it, is a state of spontaneity in which the myriad things exist in an incessant flux and transform themselves according to their own immanent principles. To life or existence is therefore attributed an essence of constant change. To the return or quest for meaning is assigned the requirement of harmony with nature; to plunge into the flux of transformation, for only within the flux can man unite himself with the infinite and enjoy the regeneration of eternal return. In other words, it is in the perfect fusion, the pre-experiential indifference of the self and nature that existence can become a meaningful

---

[20] Hightower, *The Poetry of T'ao Ch'ien*, p. 129.
[21] Hightower, *The Poetry of T'ao Ch'ien*, p. 50.

oneness. These ideas form the undercurrent of one of T'ao Ch'ien's most famous poems, the fifth of "Twenty Poems After Drinking Wine":

> "I built my cottage among the habitations of men,
> Yet there is no clamour of passing carts and horses.
> You would like to know how it could be?
> With the mind detached, one's place becomes remote.
> Picking chrysanthemums by the eastern hedge
> I catch sight of the distant southern hills:
> The mountain air is lovely as the sun sets
> And flocks of flying birds return together.
> In these things is the fundamental truth
> I would like to tell, but lack the words."[22]

The absolute detachment of the mind is a spiritual discernment derived from the soul's orientation to nature. It not only transfers the hut into remote distance and absolves the poet from worldly corruption, but enables him to perceive the epiphany of oneness between chrysanthemum, hill, and self.

In the second part of "The Return" we may distinguish several stages in the poet's redintegrative process: a linear progress home, a re-identification with its surroundings and the garden reunion, a harmony with nature, and an assimilation with cosmic change. Due to the human and existential spirit of the pastoral (and in contrast to "Encountering Sorrow") the linear progress acquires no ritual significance beyond structural transition. The rhythms of rocking boat and wafting breeze cease with the sudden appearance of home. Anxieties end as external and internal landscapes become allied. Next the alliance is extended with the thematic entrance of domestic bliss. A sense of spiritual belonging and contentment—a combination of a cup of wine, a simple window sill, and a little room—reflects the accessibility and the self-sufficiency of the garden. This in turn leads the poet to proclaim willingly the end of desire and makes his further reunion with the garden possible.

> "My boat rocks in the gentle breeze,
> Flap, flap, the wind blows my gown;
> I ask a passerby about the road ahead,
> Grudging the dimness of the light at dawn.
> Then I catch sight of my cottage-
> Filled with joy I run."[23]

Within the garden the poet's contentment fosters an assimilation with the simplicity and spontaneity of nature. This is a world of perpetual motion; clouds

---

[22] Hightower, *The Poetry of T'ao Ch'ien*, p. 140.
[23] Hightower, *The Poetry of T'ao Ch'ien*, p. 269.

and birds alike share a "homing" impulse. As a microcosm of the world of change, the garden offers reconciliation of an estrangement which the corrupt secular world could not. In the second part of the poem, the poet moves beyond his peaceful enclave to delight in the exploration of his newfound harmony with nature: "sensing my life of movement has come/to rest." Paradoxically the "rest" of the poet is within the centre of an ever-changing nature, for rest is not quiescence but active involvement and identification with natural cycles and eternal return: "nature's myriad creatures flourishing in season." Finally, reflecting upon the mundane world and the state of nature, the poet achieves a transcendent level of discernment. The new consciousness is a philosophy of self-assimilation within the Great Change; it is a renunciation of both wealth, honour, and the paradisal myths of Mount K'un Lun, and a triumph over death. Of most importance is the promise of a harmonious life of contentment for each individual without recourse to the supernatural.

From a structural viewpoint, "The Return" presents the reconciliation of man and nature through spatial and temporal orientation. Beginning with the self's re-identification with home in a single day, the poem expands its focus to a daily reunion with the garden, to a seasonal harmony with nature, and finally to total assimilation with the incessant change of the infinite. As a pastoral abstraction of the mentality behind the cyclic myth in terms of man's harmony with nature as spontaneous change and existence within an integral oneness, "The Return" offers an escape from the tragic sense of randomness, estrangement and finitude. It is unique and extremely influential in Chinese literature.

2. The Mythic and Philosophical Aspects of the Plays of *The Peony Pavilion* and *A Dream in Han-tan Inn*

a. The Metaphysical Basis of T'ang Hsien-tsu's Plays

As a mature art form Chinese drama owes much to Emperor T'ai-tsung of the Yüan dynasty (1277–1368). This Mongol emperor's distrust of scholars led to the abolition of academic examinations for those seeking official preferment. As a result, in the late thirteenth century many scholars turned to the theatre and replaced the playwright-actors as masters of the most popular art of the time. Evolving from such diverse sources as mediumistic seances, ritual and court entertainments, martial mask-dances, puppet shows and shadow plays, and of course, colloquial storytelling, the "Variety Play" of the Sung dynasty (960–1279 A.D.) had already achieved a well-integrated narrative structure. But for the most part, these Sung plays were the creations of the book guild, actors, and folk writers, and it remained for the Confucian scholars, robbed of court positions by Emperor T'ai-tsung's decree, to perfect the two mainstreams of Chinese theatre.

The Northern Play, popular in the capital and often composed by poets as an intellectual pastime, was a mature drama of four scenes' duration. Unfortunately, from a critical viewpoint, these early plays were judged not so much by plot action or character development as by the lyric poetry chanted by the two lead roles. Contemporary with the Northern Play, but dramatically more interesting, was the Southern Show of Hang-chou, the most prosperous city of the southern districts. According to Lin Wen-keng, the Southern Show was a full-fledged play of song and dialogue written in colloquial language; structurally, the dialogue advanced dramatic action, while the songs, arranged in sequence, vividly expressed the heightened sentiments of the characters.[1] By 1260 A.D., with the unification of the Yüan empire, the two dramatic forms merged, with the Southern Show characteristically dominant. With the rise of the Ming dynasty (1368-1644 A.D.), the Yüan drama was further refined into "Ch'uan-ch'i," literally "Tales of the Fantastic," or to be more specific, "Dramatic Romance," in order to distinguish it from the fictional Ch'uan-ch'i of the T'ang dynasty.

The Dramatic Romance (Ch'uan-ch'i) is perhaps the lengthiest of dramatic sub-genres, composed of forty or more scenes and characterized by songs chanted in alternate solos or in chorus by several characters in the play. The greatest dramatic achievement of the Ch'uan-ch'i genre, and probably of the entire Ming dynasty, were the "Four Dream Plays" by T'ang Hsien-tsu (1550–1616). Although contemporary with Shakespeare, the intellectual milieu of the Ming period is so utterly different from the Elizabethan that its philosophical background will be discussed before turning to the work of the dramatist himself.

In a very concrete sense the Ming thought originated in an experience of the self and an aspiration to sagehood. The single key to this was the rationalistic Neo-Confucian doctrine that man in his essential nature is identical with all nature and is of the same substance forming one body with Heaven, Earth, and all things. Theoretically this identity is based on the equation of "Jen" (human-heartiness, humanity, love) with life itself; the fundamental characteristic of the universe is its productivity or creativity, and man too is seen as creative in his very essence. The Ming thought held that self-transcendence could be attained through one's ethical and cultural activity participating with the creative process of Heaven and Earth, and by affirming one's humanity wherein one's spontaneous desire was naturally in accordance with Heaven. Thus not only did sagehood depend on one's speculation of the mind and nature of man, but man's bodily self and his moral mind were posited as the centre of creation, and the direct attainment of sagehood was justified.

---

[1] Lin Wen-keng, *History of Chinese Literature*, p. 166.

Of the many schools of Neo-Confucian thought, the Idealism of Wang Yang-ming (1472–1529) found an advocate in Lo Ju-fang (1515–1589), whose vitalistic and relatively existentialist philosophy was most adaptable to the romantic, emotional, and sensual temper of the time, and serves as the best metaphysical basis of T'ang Hsien-tsu's dramatic work.

As a prominent "existential" Confucianist of the Left Wing School, Lo Ju-fang regarded the perpetual renewal of life as a ceaseless vitality, intrinsically good, and as the animating principle of the universe. He equates "sheng" (life or vitality) with "jen" (seed or love), and further identifies this jen with "jen" (man) in the fact that the birth of a person is due to the latent vitality implicit in the process of creation and is a part of the joy of spontaneous creativity. By equating man, love, and vitality, Lo Ju-fang has existentially identified human nature with the inherently good vitality of life, and assimilated the self to the perpetual regeneration of the universe through humanity as an incessant, animating power of creation. In his philosophy man is truly the cross-roads of creation, spontaneously free, but responsible for maintaining the way of nature. This vitalistic strain of Lo Ju-fang's thought is evident in T'ang Hsien-tsu's "Four Dream Plays," especially in *Mu-tan t'ing (The Peony Pavilion)*, and *Han-tan meng chi (A Dream in Han-tan Inn)*.

b. The Peony Pavilion (*Mu-tan t'ing*): An Incessant Pursuit from Limbo to Resurrection

Despite a lifelong passion for the Yüan drama, T'ang Hsien-tsu's literary efforts did not begin until his retirement from a controversial magisterial career. Settling in Lin-ch'uan, his family home, he completed *Mu-tan t'ing huan-hun chi (The Return of the Soul to the Peony Pavilion)* in 1595. The play, usually referred to as *The Peony Pavilion*, is a romance of fifty-five scenes which demonstrates mastery of both colloquial and poetic diction (especially with regard to puns and allusions), and skilful characterization in its dramatization of human conflict within the fantasy of a dream allegory. An English translation and a study of the play are readily available by Cyril Birch and C.T. Hsia; a brief synopsis will serve here for analysis:

Following the seasonal rotation, which is one of the structural hallmarks of the play, the spring phase begins in the Southern Sung dynasty with a promising young scholar, Liu Ch'un-ch'ing, literally "Lover of Spring." As a gardener in Kuangtong, Liu dreamed of a beautiful maiden beckoning him to a life of virtuous prosperity from beneath a plum tree. Obsessed by her beauty and his Confucian desires for official success, Liu adopts a new name, "Meng-mei," literally "Dreamer of the Plum." Meanwhile, Prefect Tu, a sternly rational Confucian, has a daughter called Fair Bride (Li-niang). Perceiving only the importance of grace and virtue in womanhood, Tu maintains his daughter in virtual seclusion. And under the tutelage of Ch'en, an aged Confucian pedant,

Fair Bride is not even allowed to wander around the garden. But the return of spring has stirred the blood of youth, and Fair Bride undertakes a quest of the heart in opposition to the naive love songs of the Golden Age, the favourites of Tutor Ch'en. Encouraged by her vivacious handmaiden, Ch'un Hsiang (Spring Fragrance), Fair Bride adorns herself in spring finery and visits the family's forbidden garden. Overcome by the emotions of spring, she soon returns to her room and dreams of a young scholar who lures her into the Peony Pavilion with a willow twig and endearing words. Their dream-romance blossoms immediately but farewells are inevitable. Fair Bride awakes to profound love-sickness, but on returning to the garden pavilion she finds only a large plum tree. In despair she consoles herself by believing that after death she could at least be buried under the plum tree so as to be near her dream love.

Fair Bride's lovesickness continues, and summer brings the shocking realization that her beauty is fading. She paints a self-portrait to immortalize her loveliness. Fair Bride's illness is, of course, a cause for parental concern; but Prefect Tu and Tutor Ch'en regard it as a mere "flu," while Madam Tu employs Nun Stone Fairy to dispel the evil spirits which cause her daughter's illness. As autumn brings the fall of beauty, Fair Bride's condition worsens. And on the Mid-autumn Day, the festival of moon and lover, she is buried under the plum tree while her portrait is hidden within the Peony Pavilion. Meanwhile Prefect Tu has been appointed Tribune in Yang-chou, but before his departure he converts his mansion and garden into a memorial convent superintended by Nun Stone Fairy and Tutor Ch'en.

In the winter phase the soul of Fair Bride descends into Hades' realm, but judgement is suspended by the administrative rectification of the tenth circle, and her soul remains in Limbo. Three winters pass, and we return to the progress of young Liu. Travelling to the capital to take an official examination, Liu slips on river ice but is rescued by Tutor Ch'en and removed to the Blossom Convent for recuperation. Meanwhile Fair Bride is saved from an inauspicious transmogrification by the Flower Goddess of the Peony Pavilion who intervenes on her behalf. The Goddess' plea so moves the Judge of Metempsychosis that he grants Fair Bride her unfinished life and she returns to the garden. Spring returns, and the convalescing Liu discovers Fair Bride's self-portrait in the pavilion. He soon realizes that this is the image of his dream-love, and calls her to rejoin the upper world (though in ghost-like form), and in three successive nights she reveals the mystery of her love, death, and resurrection, bidding Liu to exhume her body that they might enjoy the fruits of corporeal love. Liu induces Nun Stone Fairy to perform the illegal deed which returns life to Fair Bride, and the elderly nun promptly marries the two lovers. The trio set out for the capital where Liu is to write his examination. Summer is a blissful period for the newlyweds, but a rebellion in the south has delayed announcement of the examination results.

94 A Comparative Study of Chinese and Western Cyclic Myths

Autumn brings an order for Tribune Tu to pacify the rebels; he sends Madam Tu, Handmaid Spring Fragrance, and the household to the capital, but is himself besieged by rebels in Huai-an. Meanwhile Tutor Ch'en, hurrying to alert the elder Tu of the scandalous exhumation, is captured by the rebels and freed on the condition that he convey to Tu false news of the death of Madam Tu and Spring Fragrance. The tragic news provides Tribune Tu with a strategy which results in victory for the imperial forces. The delighted Emperor promotes Tu to the post of Prime Minister and appoints Ch'en as the Palace Announcer. The examination results confirm Liu as the first ranked scholar, and the capital is puzzled by the unknown whereabouts of the new literary champion.

In the interim Liu has rushed to the aid of the besieged Tu, only to find himself arrested at the victory banquet and charged with illegal exhumation; naturally the rational Tu has rejected the story of Fair Bride's resurrection and Liu's claims to be his son-in-law. Meanwhile Madam Tu and Spring Fragrance on their journey to the capital have met Fair Bride in a deserted house which she was sharing with Liu. Oblivious to this ghostly reunion, Prime Minister Tu extracts a confession from Liu and sentences him to death, but he is soon rescued by imperial officials sent to find the new literary champion. Still furious, Tu brings the crime to the Emperor's attention and all are summoned to the court for imperial judgement. Fortunately Fair Bride and Madam Tu appear in time to disclose the truth of the resurrection. With the aid of a magic mirror the Emperor infers that Fair Bride is indeed alive again, and although the stubborn Tu is reluctant to accept the verdict, his rancour is soon overcome by his love for Fair Bride. The play concludes with a joyous family reunion and the reconciliation of all parties concerned.[2]

As will be shown, *The Peony Pavilion* is a revelation of man's unity with the perpetual renewal of the universe through the triumph of love over life and death, and the reconciliation of love and propriety, or rather the reunion of reason and passion. The playwright's operative device is the heroine's total devotion to love; the subject of her quest in life and dreams, the cause of her death and resurrection, and the sustaining force of her reconciliation with the cosmos. T'ang Hsien-tsu attached supreme importance to love as the distinguishing feature of human existence, thus we read in the preface to *The Peony Pavilion*:

> "Of all the girls in this world, who is ever so steadfast and committed to love as Fair Bride? Once dreaming of her lover, she falls sick; and her illness becomes worse with her ever deeper

---

[2] This synopsis of the play is adapted from T'ang Hsien-tsu, *Mu-tan t'ing* (*The Peony Pavilion*) (Taipei: Wen-kuang ch'u-pan-she, 1974). Also cf. Cyril Birch, trans., *The Peony Pavilion* (Bloomington: Indiana University Press, 1980).

attachment to love until she draws a self-portrait as a legacy to the world and then dies. Dead for three years, she can still in her limbo-like existence seek her dream-lover and regain her life. To be as Fair Bride is truly to be one totally devoted to affection. Love is of source unknown; but remaining true and totally committed to it, one may die of it and again come to life by its power. Love is not love at its fullest if one who lives is unwilling to die for it, or if it cannot restore to life one who has so died. Love engendered in a dream is not necessary to be unreal; there is no lack of such dreamers in the world. Only for those whose love must be fulfilled on the pillow, and for whom affection deepens only as old age draws on, it is entirely a corporeal matter... "[3]

In a sense, this eloquent statement of T'ang Hsien-tsu's philosophy of love echoes Lo Ju-fang's existentialistic doctrine of identification of the mind with the generating force of life, and the assimilation of the self with the incessant transformation of the cosmos. *The Peony Pavilion* is the playwright's postulation of love as the primary and essential condition of life, an affirmation of life consonant with Lo Ju-fang's philosophy that human senses and sentiments are as natural as the ultimate truth of the cosmos. T'ang Hsien-tsu postulates his drama of true affection in opposition to the decorous yet frigid life of the rationalistic Ming Confucianists. Thus he perceives the futility of the rational mind in analyzing the depths of human affairs: "Alas, affairs of the world are certainly beyond mortal man's full understanding. With no omniscience, one can only strive to use 'reason' as a guide to his understanding; and yet what is without in 'reason' is never sure to be necessarily within 'affection'."[4] He discerns that love as the truest and most spontaneous of human desires is an expression of selfhood and the core of the universe, enabling one to unify the self and things, to discern reality and illusion, even to transcend life and death.

The Peony Pavilion is undoubtedly a dramatic romance underscored by the cyclic myth, with an affectionate self at the thematic centre. In structure the play is closely correlated to the temporal progress of the seasonal cycle. It is the time of spring which provokes love's awakening in hero and heroine. The conversion of the hero's name from "lover of spring" to "dreamer of the plum" indicates the rise of desire and ambition in the heart of the flower and fruit planter after his oracular dream, while spring stirs in the heroine a physical and mental awakening, a yearning for love which neither reason nor propriety can suppress.

---

[3] Hsia C. T., "Time and the Human Condition in the Plays of T'ang Hsien-tsu," in Wm. Theodore de Bary, ed., *Self and Society in Ming Thought* (New York and London: Columbia University Press, 1970), p. 276.

[4] *Ibid.*

Her foray into the forbidden garden, symbolic of her union with nature, marks the beginning of a search for selfhood in contact with nature; sense and sentiment awaken through the quest for love compelled by her revelational dream. With the end of spring her love is unfulfilled in reality and paralysed by decorum, resulting in her extreme love-sickness.

While summer is the time when we would expect love to achieve fruition, Fair Bride's romantic difficulties create a tension between actuality and the seasonal expectation. This is reflected in her self-portrait, a static image of love, unfulfilled yet never changing. It is her legacy to the world. The tension is maintained through autumn when love should bear harvest. The disparity between the natural pattern of love which has been symbolically linked to seasonal progress, and the barren reality of the heroine's love-sickness is underscored by her declining health and eventually, death.

The descent of Fair Bride's soul into Hades and its three-year imprisonment in Limbo is of course thematically consistent with the attributes of winter, the time of death and hibernation. With the return of spring, love and health are reborn in hero and heroine. Resurrection and reunion lead to marriage, and as love is given corporeality, the play's thematic action is again in harmony with the seasonal cycle.

With harmony re-established, summer corresponds to marriage and maturity, and autumn is allowed to generate its harvest of reunions progressing from dream to reality. Thus the mother/daughter reunion is undertaken through the former's ghost-like state of uncertainty and the latter's death-like status of fugitive from the southern rebellion; the daughter/father reunion is fulfilled during the former's death-like faint, while the father/son-in-law reconciliation is completed only after the latter's literary success and an imperial verdict.

Indeed, the temporal progress and thematic structure of the play are in perfect accordance with the attributes of the Chinese cyclic myth: spring, birth or resurrection; summer, growth or attachment; autumn, ripeness or harvest; and winter, death and the potentiality of rebirth. It is even analogous to the symbolic forms of the seasonal force in King Wen's Eight Trigrams: spring, the arousing; summer, the clinging; autumn, the joyous; and winter, the abysmal.

Thematically speaking, the correlation of the play and the cyclic myth is furthered by the possibilities for transcendence in dream and death. To recall that at the heart of the myth is the relief of temporal tensions through union with the natural cycle, is also to suggest the difficulty of the union within the linear constructs of human consciousness. Whether dreams are linear in the sense of recollection, or spatial and cyclical in the sense of Carl Jung's "synchronicity," they certainly allow, a freedom of symbolic self-identification usually denied in waking consciousness. The dreams in *The Peony Pavilion* are both naive and oracular; they are engendered through the unconscious wish-fulfilment of the unsatisfied self and are immediately the motives for the hero and heroine's incessant quests for love. The transcendability of dream over both time and the strict decorum of the censorious ego free the heroine from all inhibitions and taboos to experience the fulfilment of love which through her dream is the only reality beyond past, present, and future.

Death and death-like faints in the play are similarly revealing. They are spontaneous, sleep-like deaths, circumscribed by the persistent will of the questing heroine. Again the transcendability of death over time, decorum, and the rational enables the heroine as "a sleeping beauty" to be inanimate for three years; yet spiritually she roams the world in quest of her love. Here death simplifies love into true humanity, and purifies love to an absolute will, a vitality to die for, resurrect, and reconcile.

While the resurrection of Fair Bride is the material demonstration of the power of love, it is above all the poetic truth of Lo Ju-fang's equation of love and humanity to life and vitality. The marriage is not only a union of questing selves but a fulfilment of perfect selfhood through love as the truest humanity and the essence of the self's unity with the perpetual regeneration of the universe.

From a philosophical viewpoint the real triumph of *The Peony Pavilion* is the perfect embodiment of the conflict between the heroine and her father; in other words the metaphysical dispute of the primacy of "love" over "reason." Tribune Tu is a stern guardian of Confucian morality. His dedication to "reason" is a block to spontaneous affection and keeps him, as official rectifier of the course of spring and agriculture, remote from the actual soil the peasant tills or the back garden where nature prevails. Although the ritual aspects of the cyclic myth are so deeply engrained in Chinese culture that even such a rationalist as Tribune Tu is officially responsible for maintaining the harmony of Tao and agricultural practices, it is this same Confucian rationalism which negates the spiritual nourishment derived from such practices. Incapable of realizing the heroine's physical and mental awakening, he refuses to acknowledge his daughter's resurrection and marriage until the very last scene of reunion.

Fair Bride as the incarnation of love initially rejects "reason" which suppresses her awakening to true selfhood and denies her subsequent personal fulfilment. Eventually she asserts her own desire and makes the supreme commitment to love in the decorum-free states of dream and death. Her incessant quest for love is first rewarded by Tutor Ch'en's conversion from a rational pedant to a naive interpreter of love songs in *The Book of Odes*. The conversion is completed as he becomes the guardian of the heroine's grave and her champion against her father's rigidity. But the ultimate revelation of love's supremacy over reason occurs in the great reconciliation of the finale. Her love, fulfilled in marriage, is sustained despite the lack of paternal approval until Tribune Tu's intransigence necessitates further action. She faints away in

Commanding nearby spirits and animals to intrigue a dream of a materially successful life, the Immortal lulls Lu Sheng into his dream quest.

The pillow is hollow, and the hole within may be identified as the birthplace of subconscious desires, dream and death; even the womb and rebirth. Life in the dream world of the pillow suggests both wish-fulfilment and ordeal, for Lu Sheng completes a libidinal marriage to Miss Ts'ui (as the earth-mother image and the goddess of wealth and fame), and endures the tests of extreme suffering typical of an initiation before enjoying the great accomplishments of the questing hero. Indeed the dream consists of extremes of glory and disgrace, blessing and disaster, and happiness and sorrow over a span of sixty years. This phase of transfiguration ends with the final reconciliation of the hero and the emperor; the atonement of the son with the father, which brings the hero to penultimate success and glory as a noble and contented lord. The return is a great awakening. Having enjoyed great wealth, fame, and longevity, the hero leaves his phantasmagoric life without regret at the age of eighty, only to awake in this world as a farmer in a torn sheepskin jacket in Han Tan Inn. This disparity between the two lives is the immediate cause of enlightenment. Having fulfilled his desires for sensuality, wealth, and fame in the transfigurative dream world, and realizing in retrospect the nature and truth of a life of sentiments, the hero is reconciled with the Taoist Immortal. Atonement of the hero and the emperor (representative of father and sage images) is also the reconciliation of the junior and senior elements of Lu Sheng's own character. This apotheosis enables the young man to become master of the two worlds and bestows on him the freedom to live. The conversion of a Confucian malcontent to a wandering Taoist symbolizes a spiritual transcendence, an incessant pursuit for pure being, and even the cultivation necessary for the harmonious unity of self and cosmos.

Thematically the dream is apocalyptic. The intransigence of life is a central metaphorical concern, illustrated by the duration of the dream; despite sixty years of extreme success and imperial service in dream-time, in reality the duration was only the time required to cook a bowl of millet porridge. This phantasmal nature brings about a disillusionment with life whether it is regarded as sentimental or rational. The dream's origin in the Immortal's intrigue and the vagueness of the hero's free will undermined by the Immortal's looming dictatorial shadow serve to undercut the hero as a legitimate seeker of self-fulfilment. The hero's dream life then must be viewed not as a paradigm of the virtuous life but as a means to an end, bringing him the greatest possible contrast to ensure enlightenment. It would appear that the fuller the dream life, the greater the contentment, and the more effective the enlightenment; for as T'ang Hsien-tsu writes, "where dream ends, awakening begins; and when affection is spent, enlightenment follows."[8] It is in this connection that it may

---

[8] *T'ang Hsien-tsu chi*, p. 1096.

be said that T'ang Hsien-tsu did not suffer a radical change of attitude which induced him in his last play to look down upon "love" or any other human attachment; nor is *A Dream in Han-tan Inn* to be regarded as a worldly renunciation of, or escapism from, the time-space world of sentiment. Rather, in the context that the truest humanity is the innermost affection, the core of selfhood, *A Dream in Han-tan Inn* is to be understood as another level of T'ang Hsien-tsu's metaphysical speculation complementary to that of *The Peony Pavilion*. Here he perceives that the attainment of pure being lies in the total detachment of self from the world of sentiment consequent to the fullest experience of that sentimental life.

While mental detachment and unity of the self with the ultimate origin both require a true selfhood defined through T'ang Hsien-tsu's philosophy of love and being in *The Peony Pavilion*, the dream paradoxically serves as both the vehicle for the fullest experience of the life of sentiment and a metaphor of the transience and emptiness of life. At the conclusion of the play, vehicle and metaphor combine into a higher spiritual realization rather than a renunciation of life. The process is analogous to the "sudden enlightenment" of the Ch'an or Zen School of Buddhism which professes the immediate and direct attainment of Buddhahood by anyone at any time and in any place. Perhaps the oracle is to be understood in a Ch'an Buddhist sense: "To cultivate (sweep) till there will be no things or no object (no petals), no self and no subject (ground), and no means and no consciousness (no broom) to become a Buddha (a sage—one who is first with and then without selfhood) to join the ultimate origin." Thus it seems clear that any perception of negativism is undermined by the playwright's existentialist vision of the fusion of Taoist myth, the Confucian ethic of sentiments, and the Ch'an Buddhist oracle.

In comparing *The Peony Pavilion* to *A Dream in Han-tan Inn* we have noted that the former demonstrates the path to true selfhood while the latter concerns the path to enlightenment or sagehood once true selfhood has been attained. With regard to the representation of the cyclic myth both plays are united in their concern for the transcendence of linear human consciousness. Such temporal concerns are indeed the limitation of the human condition and a perennial subject in Chinese literature. While the seasonal pattern in *The Peony Pavilion* is perhaps more representative of the survival of periodic regeneration in Chinese drama, *A Dream in Han-tan Inn* demonstrates an affinity in terms of ultimate unity. Where the spontaneous expression of love as human nature necessitates a harmony with the natural cycle, that harmony is in turn transcended by the ultimate and original unity. That the drama of T'ang Hsien-tsu should provide such a model of transcendence in the human affairs of literature is one of the great achievements of the Ming culture.

3. The Mythic and Philosophical Aspects of the Novels of *The Journey to the West* and *The Dream of the Red Chamber*

a. The Historical Background

Originating in the philosophical and historical writings of the earliest Chinese classics, Chinese fiction first flourished in the fourth to sixth centuries A.D. With the popularity of the supernatural tales common to Taoist and Buddhist religions, these shorter narratives often focused on myths, legends, and tales of the fantastic, although conversational pieces regarding ancient and contemporary celebrities were not unknown. But it was not until the T'ang dynasty (618–906 A.D.) with the introduction of the Ch'uan-ch'i (Tales of the Fantastic) genre that the narrative form shed its poetic excesses and interpolations, and evolved an effective style based on less-adorned prose and more frequently depicting tales of the marvellous. The stories of the T'ang dynasty paved the way for the vast scope of Sung fiction and the immense popularity of its practitioners. In the Sung dynasty (960–1279 A.D.) storytelling became an accepted profession, each bard specializing in one of four areas: tales of chivalry (often military) and litigation; the religious story popular in Buddhist sects; the historical recitation; and the realistic love-story often coupled with proficiency in tales of the supernatural. In the hands of Sung storytellers historical materials and literary fragments were expanded into complete stories, rich in detail and vivid description. This oral tradition was further refined and enriched by the Yüan dramatists in their plays. Thus there evolved in the course of time immensely popular story cycles which provided the subject matter for many of the earliest Chinese novels. By the last century of the Ming dynasty, fiction had become an established literary form among scholars and the populous middle class of the greater urban areas. With a sophisticated audience the colloquial short story reached the peak of its development with the publication of numerous collections in the first decades of the seventeenth century. Meanwhile the Chinese novel was entering its "golden age," a period extending into the early twentieth century.[1]

Of the bulk of Chinese fiction in the past several centuries, six novels stand out as representatives of the "story cycle" folk tradition or of an individual author's creative genius: *The Romance of the Three Kingdoms*, *The Water Margin*, *The Gold-vase Plum*, *The Scholars*, *The Journey to the West*, and *The Dream of the Red Chamber*. Of these works, the latter two are the most illustrative of aspects of the cyclic myth.

---

[1] For a literary and historical background of Chinese fiction, cf. Liu Wu-chi, *An Introduction to Chinese Literature* (Bloomington and London: Indiana University Press, 1966), pp. 141–228.

b. The Journey to the West (Hsi-yu chi): A Celestial Progress to Divinity

Hsi-yu chi, known to English readers as **Monkey** through Arthur Waley's abridged translation of 1942, or as *The Journey to the West* through Anthony Yu's translation of 1977 and W. J. F. Jenner's translation of 1982, is a combined product of the literary cycle of oral tradition and the author's creative imagination. Through its five hundred years of evolution it had existed as a crude colloquial story, a poetic novella, and a six-part drama before Wu Ch'eng-en (1500–1582) refined it into its present novelistic form. As scholastic studies of the novel are available by C. T. Hsia, Glen Dudbridge, and James S. Fu, we will confine ourselves to the aspect of the cyclic myth in the novel.[2] Generally *The Journey to the West* consists of four parts: (1) a myth of the divine birth, quest, and ultimate defeat of a stone monkey; (2) a pseudo-historical account of Monk Tripitaka's Oedipus-like life story and family reunion, before his pilgrimage to the Holy Mountain of Buddha; (3) a journey of eighty-one ordeals undertaken by Tripitaka and his three disciples Monkey, Pigsy, and Sandy; and (4) the ultimate reconciliation of the pilgrims with the Infinite Buddha. While the plot of the novel is wondrously complex, a general understanding of major incidents is prerequisite to understanding its structural basis within the cyclic myth. With this end in mind, we offer the following summary:

At the outset of the novel it is apparent that Monkey's miraculous birth has resulted in an innate knowledge of cosmic harmony, for he sprang to life from the rock peak of the holy Mount Flower and Fruit which for centuries had received the spirits of Heaven and Earth, and the sun and moon. Thus the hero's first action was a reverential bow to each of the four quarters of the universe. With his supernatural talents and abilities Monkey became King of monkeys, gibbons, and baboons, and ruled for several hundred years in the contentment of natural harmony. But a sudden awareness of life's transience and death's inevitability compelled Monkey to search for Buddhahood, the enlightenment which denies the dark kingdom of death. Following an eight-year quest, Monkey arrived on the Western Continent and under the tutelage of a woodcutter was

---

[2] Biographical details about Wu Ch'eng-en (1500–1582) are sketchy, but he is believed to have been a native of Huai-an in Kiangsu province, and to have had no successful official career except as a minor district magistrate for seven years, despite the fact that he was a learned Confucian enjoying a wide reputation for his wit and literary talent. He is said to have followed the existing story cycle of folk tradition to compose *The Journey to the West* at the age of sixty-eight. A volume of classical poetry and prose left by him has been edited by Hsiu-yeh Liu, with valuable biographical and bibliographical information. See Liu Shiu-yeh, *Wu Ch'eng-en shih-wen chi* (Shanghai: Ku wen-hsüeh ch'u-pan-she, 1958). Also cf. James S. Fu, *Mythic and Comic Aspects of the Quest* (Singapore: Singapore University Press, 1977).

accepted as a Buddhist novitiate, and given the religious name, "Aware-of-vacuity." Following seven years of study, Monkey was rewarded with the secrets of longevity and seventy-two magical transformations of which levitation over great distances proved to be the most important. Forbidden to mention his discipleship upon punishment of death, the hero left the Western Continent.

Returning to his homeland in time to free his subjects from the devastations of the Demon of Havoc, Monkey united his "people" into an invincible army and was declared the greatest king of the animal and demonic worlds. Searching for a weapon to befit his new status, he obtained from the Dragon King of the Eastern Sea an iron bar, "The Golden-clasped Wishing Staff." The mighty weapon possessed cosmological significance as the leveller of ocean and river bottoms, and as the instrument used to fix the course of the Milky Way. With this magical staff at hand the hero accepted armour from rulers of the remaining three seas and returned to Mount Flower and Fruit.

After a great banquet Monkey was (in a dream) carried by demons to the edge of the City of Darkness, but he awoke and recklessly confronted the Ten Judges of Death, demanding immortality for himself and all his subjects. Fearing that the granting of such a request would upset the harmony of Light and Dark, the Dragon King and the First Judge appealed to the Supremacy of Heaven to arrest Monkey. This military option was disregarded; instead Monkey was offered the immortal post of Supervisor of the Celestial Stables which he accepted. Fifteen years passed before he was aware of the relatively low rank of his position, and in a rage he returned to his terrestrial kingdom. Thereupon a celestial army of spirits and gods was dispatched to arrest Monkey, but they were terribly defeated. In an effort to restore peace he was offered a new post, the Great Sage in Equal Of Heaven. Monkey accepted and then returned to the celestial realm.

Our hero's appointment in Heaven was typified by several instances of reckless greed; guarding the Peach Trees of Immortal Divinity led him to devour many of the sacred fruits, and after intriguing his way into a celestial banquet, he consumed all the divine nectars and ambrosia laid out for the feast. Intoxicated on the fruits of his last misdeed, Monkey entered the Palace of Lao Tzu, and swallowed all five gourds of the elixir of life. Realizing his guilt he fled from Heaven to the securities and honours of his lower world.

Monkey's escapades so enraged the Supremacy of Heaven that all the forces of the universe were commanded against the mischievous protagonist, and he was finally defeated by the Diamond Snare of Lao Tzu and sentenced to death. Immune to the usual weapons and thunderbolts, Monkey was imprisoned in Lao Tzu's Crucible of the Eight Trigrams but he survived the alchemic fires through his knowledge of the nature and structure of The Eight Trigrams, and escaped to battle his way to the very doorstep of the Supremacy of Heaven. Finally the Buddha himself intervened and Monkey was tricked into a moment of bewilderment; a moment which allowed Buddha to press and imprison him under the cosmic mountain, Mount Five Elements, to do penance until rescued.

The ultimate defeat of Monkey concludes the first section of the novel, and the incidental detail provided should suggest the encyclopedic nature of its cosmological references. Borrowing freely from the myths of antiquity, the religious lore of Buddhism and Taoism, and folklore of the original "stone monkey" story cycle, Wu Ch'eng-en has fashioned a work which entertains as pure adventure, and instructs as religious allegory. With this cosmological framework in mind, and for brevity's sake, further events of the novel may be treated with less regard to incidental detail.

The second major section of the novel concerns the Oedipus-like biography of monk Tripitaka. Ch'en Kuang-jui, the new literary champion of the Academic Examinations, was awarded by an official post in the distant River Province and the hand of the Prime Minister's daughter. As the newly-weds were travelling towards their new home, Ch'en Kuang-jui was murdered by a jealous ferryman, Liu, who then posed as the newly-appointed governor and obtained silence from Lady Ch'en, who was anxious to protect her unborn child. When her son was born an oracular advice prompted Lady Ch'en to brand the infant and send him downstream on a wooden plank with a letter of identification. The baby was discovered by an abbot and raised in a Buddhist temple, and at age seventeen was named Hsüan Tsang, literally "Great Obscurity." Setting off in quest of his origin, Hsüan Tsang eventually discovered the truth of his birth and deposed the imposter. The grand reunion is completed with the appearance of his father, Ch'en Kuang-jui, who had been resurrected through divine intervention.

In the meantime, events had occurred to set the stage for the third section of the novel. Emperor T'ai-tsung of the T'ang dynasty was once summoned to be a witness by the Supreme Judge of the Dark Kingdom. He descended to Hades in a death-like dream and perceived the manifold suffering of those in the underworld. upon his resurrection Emperor T'ai-tsung appointed Hsüan Tsang to preside over an Imperial Grand Mass for the Dead. Meanwhile the Buddhisattva of Mercy had been searching for a pious believer to deliver the sacred Scripture of Tripitaka to the lustful and evil inhabitants of the Southern Continent. Inspecting the route the pilgrim would take, the Buddhisattva converted two monstrous incarnations to Buddhism: "Sandy" (in religion "Aware-of-purity"); and "Pigsy" (in religion "Aware-of ability"); and bid them to assist the pilgrim's passage and offer assistance. In addition she arranged a pardon for the Dragon Prince of the Western Sea and transformed him into a white horse to bear the future pilgrim; and finally, Monkey was converted to Buddhism, re-named "Aware-of-vacuity" and ordered to await the pilgrim's arrival when he would be released from Mount Five Elements to protect the pilgrim's westward progress. Eventually the Buddhisattva travelled to the site of the Imperial Grand Mass and invested Hsüan Tsang as the bearer of the Great

Vehicle of the Scripture of Tripitaka. Henceforth, Hsüan Tsang the pilgrim is to be known as Tripitaka, and the journey begins.

In the novel's third section Tripitaka was joined by Monkey, Pigsy, Sandy, and the Dragon Horse. They then encountered Zen Master Crow-nest who bestowed upon them the scriptures of the Heart Sutra as a spiritual companion on their journey. For the next fourteen years, through one hundred and eight thousand leagues of travel, the five pilgrims endured eighty-one divine and demonic ordeals. Finally they crossed the Sacred Water surrounding the Holy Mountain of the Western Paradise and the Buddha of the Infinite rewarded them with the five thousand and forty-eight scrolls of the Great Vehicle of the True Tripitaka. The travellers and scrolls were escorted back to China on a puff of fragrant wind by the eight Vajrapanis (Diamond Angels), and upon their subsequent return to the Western Paradise, each was rewarded through various celestial appointments; as Buddhas in the cases of Monkey and Tripitaka, or as holy guardians in the cases of Pigsy, Sandy and Dragon Horse. In this manner is the fourth and final section concluded, and the great reconciliation fulfilled.[3]

Quest and progress form the structural framework of *The Journey to the West*. We find Monkey's universal pursuit of immortality and Tripitaka's persistent search for his origins both secular and divine. Moreover the journey concerns spatial progress from the terrestrial to the celestial, and temporal progress (in terms suggested by Mircea Eliade in *Cosmos and History*) from "profane time" to "great time" and finally to "no time," the time of reconciliation and the union of the pilgrims with the Infinite. In a spatial sense the novel is a divine comedy between the Heaven of celestial deities under the Jade Emperor of Taoism, the Western Paradise of Buddhist saints and arhats headed by the Buddha of the Infinite, the Underworld of Darkness where ghosts and souls are ruled by the Ten Judges of Death, the demonic world of monsters and goblins, and lastly, the terrestrial world of earthbound mortals.

Thematically speaking, whether the novel is regarded as a religious or philosophical allegory, a mock epic, or even as a revolutionary satire of decadent bureaucracy, one cannot escape the surviving threads of the cyclic myth which provide the cosmogonic background and the individual impulse to quest. The temporal tensions which provoke Monkey's quest are no different from those experienced by Titan K'ua Fu several millennia before in the time of myth; what is different is Monkey's knowledge of cosmological harmony, the product of centuries of abstraction and evolution of the original cyclic myth.

As a religious allegory the explicitly Buddhist references in the novel are tempered by the prevailing syncretic spirit of Chinese philosophy; it must be

---

[3] For the summary, cf. Wu Ch'eng-en, *Hsi-yu chi* (Tainan: Wang-chia ch'u-pan-she, 1975); Arthur Waley, trans., *Monkey* (London: Penguin Books, 1973).

remembered that *The Journey to the West* was composed from an oral tradition of story cycle in a time when idealistic Neo-Confucianism was dominant in Chinese life. The thought of Wang Yang-ming maintains that the mind is the universe in which things are nothing but what the mind determines to realize, and in which sagehood lies immediately attainable within one's own nature. It interpenetrates that of the school of Mahayana Buddhism so dear to the pilgrim, Hsüan Tsang. This emphasis on the primacy of mind had been noted in such traditional commentaries as Ch'en Yüan-chih's preface to a sixteenth century edition of the novel:

> "There was an old preface which I read through.... It held that Monkey was symbolizing the spirit of the mind, the horse was symbolizing the coursing of the desire or the will, and Pigsy was equivalent to the eight physical desires.... As to monsters and demons, they were obstruction and illusion created by mouth, ears, nose, tongue, body, mind, fear, distortion and fantasy. They were given birth by the mind and were submissive to the mind. Therefore, in order to return to the Ultimate Origin where the mind yields to no allure or illusion, one has to regulate the mind to subdue monsters and demons to restore truth."[4]

Indeed, the symbolic expression "Monkey of the Mind and Horse of the Will" recur within the system of allusive verse chapter-headings and at large as an allegorical device within the text. Originating in the translated Buddhist sutras, Monkey and Horse symbolise the indulgent waywardness of the human mind before it attains the state of quietude and composure through Buddhist disciplines. This extensive though superficial metaphor is subsumed within the novel's central symbolism, that of the Heart Sutra.

As the central text of Mahayana Buddhism, the Heart Sutra is duly recorded in Hsüan Tsang's own standard translation.[5] Its basic teaching is that form is emptiness and emptiness is form; only through the utmost vacuity of the senses and sentiments is it possible for one to access Nirvana. Early on his journey, Tripitaka encounters Zen Master Crow-nest, who bestows on him the Heart Sutra as a spiritual companion to protect him on his perilous journey. But in Buddhist allegory the Heart Sutra is a far more important guide for Tripitaka than his monster-disciples; a true understanding of its teachings would automatically expose the illusory nature of his calamities and hence reduce the

---

[4] Cf. Glen Dudbridge, *The Hsi Yu Chi: A Study of Antecedents to the Sixteenth Century Chinese Novel* (Cambridge University Press, 1970), p. 174.

[5] Cf. Hsia C. T. and Hsia T. A., "New Perspectives on Two Ming Novels: 'Hsi Yu Chi' and 'Hsi Yu Pu'," in Chow Tse-tsung, ed., *Wen Lin: Studies in the Chinese Humanities* (Madison & London: University of Wisconsin Press, 1968), p. 235.

temporal and spatial distance to Paradise. In fact such knowledge would render unnecessary the service of his disciples, for as Monkey demonstrates, all monsters and demons are both created by and submissive to the mind.

> "When the monks discussed the tenets of Buddhism and the purpose of the pilgrimage... Tripitaka remained silent, pointing at his heart and nodding again and again. The monks did not understand him... and he said: 'It is the mind that gives birth to monsters of every kind, and when the mind is at rest they disappear.'"[6]

Tripitaka is the embodiment of the fearful self-consciousness of an everyman enslaved by the senses, by humanitarian affections, and by illusory external phenomena; he is too obsessed with the phenomenal being to perceive fully the transcendental meaning of the Heart Sutra and thus rout the terrors of the senses or detach himself from delusion. Every calamity which befalls him demonstrates anew his incomprehension of the philosophical issues at hand. On the other hand Monkey is "Aware-of-vacuity"; he is the only one who comprehends the doctrine of emptiness and this accounts for his first action upon joining Tripitaka: the slaying of the six thieves of Eye, Ear, Nose, Tongue, Mind, and Body, an allegorical event indicative of his superior spiritual detachment. With his knowledge of the Heart Sutra, Monkey is able to discern various demonic delusions, is capable of numerous transformations, and free to roam the celestial and sub-terrestrial realms. He continually reminds his master to heed the teachings of the Sutra:

> "Reverend master, you have forgotten the verse, 'No eye, ear, nose, tongue, body and mind.' Of all of us who have forsaken the world, our eyes should not see colour, our ears should not hear sound, our nose should not smell, our tongue should not taste, our body should not feel cold and heat, and our mind should not harbour vain illusions: this is known as 'routing the six thieves.' Now your mind is constantly occupied with the task of seeking the scriptures, you are afraid of the monsters and unwilling to give up your body, you beg for food and move your tongue, you are fond of sweet smells and titillate your nose, you listen to sounds and excite your ear, you see things around you and strain your pupils. Since you have welcomed these six thieves on your own invitation, how could you hope to see Buddha in the Western Paradise?"[7]

---

[6] Lu Hsün, *A Brief History of Chinese Fiction*, trans., Yang Hsien-yi and Gladys Yang (Peking: Foreign Language Press, 1958), p. 218. Also cf. Arthur Waley, *Monkey*, p. 136; Wu Ch'eng-en, *Hsi-yu chi*, p. 86.

[7] Hsia C. T., *The Classic Chinese Novel* (New York and London: Columbia University Press, 1968), p. 129. Also cf. Wu Ch'eng-en, *Hsi-yu chi*, p. 297.

The thematic necessity of a full narration of eighty-one ordeals maintains Tripitaka's spiritual blindness for much of the novel. Fortunately the possibility of immediate enlightenment is one of the central tenets of Zen or Ch'an Buddhism, and Tripitaka is enlightened while crossing the Sacred Water to the Western Paradise. Spiritual transcendence is further indicated by the pilgrims' rapid and magical return to their homeland.

> "In speaking of the return to homeland on a puff of fragrant wind it describes the ease with which the True Way may be attained. If men could with their power of sight first see through the affairs of the world, then suppress the Monkey of the Mind and bind the Horse of the Will, and again with wisdom govern their anger and subdue all evil spirits—what difficulty would there be in attaining the Way?"[8]

Certainly it would be inadequate to propose that *The Journey to the West* is merely a philosophical commentary on the Heart Sutra, and while the search for Nirvana is perfectly analogous to the transcendence of temporal tensions (the original motive of the cyclic myth), we shall find more complete correlations in the novel's analogic and mythic aspects. In the analogical aspect, the food conceit bears in the novel a testimony to the primitive view of food as mana, possessing a "spirit" which connects the divine and the profane, and confers special powers upon the partaker. Such sacred foods as the fruits of the Peach Garden and the various ambrosia and elixirs of Paradise seem to be "distillations" of the regenerative force which drives the seasonal cycles. It is no accident that in his search for immortality, many of Monkey's pranks involve the eating of such fruits and nectars. Much of the novel's initiating plot concerns the repercussions for universal harmony of a mere terrestrial gaining access to such potent "mana." To restore the cosmic balance Monkey must be exiled from Heaven, and his quest for reconciliation with the divine must begin anew. In a more bizarre sense, the pilgrims themselves, especially Tripitaka, by virtue of their higher knowledge become "mana" objects to their aggressors. Tripitaka must endure perpetual cannibalistic and sexual assaults by the male and female monsters who believe that to eat his flesh or absorb his semen is to acquire the precious gift of immortality. The monsters' ferocious aggression upon Tripitaka and Monkey's incessant fight for the Fruit of Life in this view depict a vigorous quest for reconciliation and unity with the eternal divinity which has been the main theme of the cyclic myth since the beginning of time.

In the mythical aspects, the novel transcends a narrow religious application by the author's intentional fusion of Confucianism and Buddhism. Not only do

---

[8] Cf. Yang T'i's preface to *Tsa-chü tung-t'ien hsüan-chi* (dated 1542), cited by Dudbridge, p. 173.

we find in Paradise geographical correlates to the Heavens of the three religious sects, but the Holy Mountain of Buddha is no less than the Cosmic Mount K'un Lun surrounded by circles of obstacles according to ancient Chinese mythology. In a mythic sense the "journey to the west" seems to be less a Buddhist pilgrimage to India than a mythic progress to paradise or the cosmic mountain; a quest for divinity and ultimate origin. Similarly the canonization of the pilgrims seems to be less a Buddhist approval than a heavenly sanction of their immortality.

Furthermore the previous states of the protagonists as exiled gods and banished immortals in penance on earth reinforces a mythic interpretation. On the one hand they suggest such Western mythical heroes as Prometheus, Oedipus, Moses, and Faust in their defiance of all authority and their quest for knowledge; on the other hand they remind us of such Chinese mythical figures as the Titans Ch'ih Yu, K'ua Fu, P'eng Tsu, and even the Archer God Hou I with his shooting down of the nine suns, his exile to earth, and his ascent of Mount K'un Lun in search of the Pill of Divinity. In this connection the tale of Monkey's conscious upward striving from inanimate stone to animal shape with human intelligence, to the highest spiritual attainment is a typical embodiment of the fundamental subjects of the cyclic myth: the mortal self's bitter awareness of the ephemerality of life, its incessant quest for immortality through rebellion against divinity, and its eventual compromise and due reconciliation with the infinite. In its mythic implications the story of Monkey is not far from the Ceremony of Chiao as a rite correspondent to the cosmic rhythm of eternal return, of the Feng Shan ritual of the emperor's thanksgiving sacrifice to the cosmic mountain as an imperial journey to the ultimate origin, or even the institution of Ming T'ang as terrestrial participation in the universal harmony.

Whereas *The Peony Pavilion* of T'ang Hsien-tsu emphasizes an existential approach to the attainment of true selfhood through a life of utmost sense and sentiment in contact with the phenomenal world, Wu Ch'eng-en prefers a merry tone of Rabelaisian mockery. Monkey is the image of the liveliest spirit of detachment. Wu Ch'eng-en values this spirit above the cannibalism of the monsters (an image of the incessant craving for immortality), or the agonizing attachment of Pigsy as the image of the grosser sensual life, or even the spiritual blindness of Tripitaka despite his everyman quality of human compassion. Monkey is the comic image of man's intelligence, and the philosophical image of the doctrine of emptiness, capable of viewing himself and both terrestrial and celestial worlds in the humorous light of his cosmic transcendence. In a word, according to Northrop Frye's theories, *The Journey to the West* is perfectly a "comic story in the mythic mode."[9]

---

[9] Cf. Northrop Frye, *Anatomy of Criticism* (Princeton, New Jersey: Princeton University Press, 1957), pp. 33–51.

c. *The Dream of the Red Chamber (Hung-lou meng)*: A Terrestrial Journey to Disillusionment

Our second novel for study completes the transition in the development of the Chinese novel from collective to individual authorship, and is regarded by many as the culmination of that development. Again referring to Northrop Frye's theories of classification, if *The Journey to the West* is to be considered a comic story in the mythic mode, *The Dream of the Red Chamber* is then a tragic story in the high mimetic mode. Its author, Ts'ao Hsüeh-ch'in (c. 1715–1763), has created a high tragedy of affection and self-realization concerning the passage of the hero from devoted attachment to self-conscious grief and eventually to bitter enlightenment and spiritual transcendence. Whether oriental or occidental, few works are its peer in the vastness of its length of one hundred and twenty chapters, the vividness of its narration over an encyclopedic range of description covering every aspect of Chinese custom and culture, and the subtlety of its portrayal of more than four hundred characters. While the novel's great length and complexity prohibit a narrative summary, it is one of the most popular of Chinese novels in translation and some familiarity assumed on the part of the reader. Here a brief outline is sufficient for our purpose:[10]

Once before there was time, when the Supernal Mother Goddess Nü Wa was repairing the sky damaged in the cosmic war between God of Water and God of Fire, she melted a great quantity of rocks on the Fathomless Crags of Mount Chaos and moulded the amalgam into 306501 large blocks to mend the broken dome. The Supernal Mother Goddess used all the stone blocks except a single odd one which was sentimental and, therefore, inadequate for the eternal celestial dome. She discarded it at the foot of Peak Greensickness, the Peak of Sentimental Sickness. Then there came time after the estrangement of Earth

---

[10] Few biographical facts about Ts'ao Hsüeh-ch'in are available, except that he was the scion of a wealthy Manchu family which in his early youth became impoverished through political reverses. At the age of thirteen, Ts'ao Hsüeh-ch'in moved to Peking with his parents who, despite reduced circumstances, maintained their connections with the Manchu aristocracy. With the ascent of Emperor Ch'ien-lung in 1736, the family briefly regained favour, but by 1744 unknown disasters had again ensnared Ts'ao in dire poverty. It was during this period that Ts'ao began composing his reflections upon his aristocratic youth and philosophical speculations into the enormous tapestry of "The Story of the Stone," or *The Dream of the Red Chamber*. Completion of the work occupied the next ten years, and he died in February, 1763. Also cf. Bing C. Chan, *The Author of the Dream of the Red Chamber* (Hong Kong: Lien-he ch'u-pan-she, 1986). For English translation, the novel finds the following editions: David Hawkes, trans., *The Story of the Stone* (London: Penguin Books, 1973); Wang Chi-chen, trans., *The Dream of the Red Chamber* (New York: Pantheon Books, 1958); John Minford, trans., *The Story of the Stone* (Bloomington: Indiana University Press, 1979); Jeanne Knoerle, *The Dream of the Red Chamber: A Critical Study* (Bloomington & London: Indiana University Press, 1972).

from Heaven. Time elapsed. Aeons passed. And the Stone, being moulded by the Supernal Goddess and thus possessing the magic power to move about and take shape at will, wandered all over the universe until he became a courtier in the Palace of Goddess of Disillusionment. He spent a lot of time idling along the banks of the Sacred Water surrounding the Western Paradise of Buddha. And there, by the Rock of Rebirth, he found the beautiful Crimson-pearl Flower. He was so fond of the flower that he watered it with sacred celestial dew till it shed the vegetable shape and assumed the form of a fairy. The Fairy, extremely conscious about her debt of gratitude to the Stone, decided to repay him with tears in a mortal lifetime, when they would both incarnate in the world below. As a consequence, Goddess of Disillusionment ordered Buddhist Fathomlesso and Taoist Mysterioso to escort Stone and Crimson-pearl Flower, accompanied by a group of passionate fairies and amorous spirits, to descend to the mundane world to be born in the clans of Duke of Glory and Duke of Peace to experience the illusory life on earth.

Meanwhile, in the city of Su-chou, next to the Gourd Temple, there lived a gentleman called Chen Shih-yin, who led a blissful life with all the domestic comforts and pleasure. One summer day, Chen felt asleep in his study. In his dream he came across a Buddhist monk and a Taoist priest and overheard their conversations about the story of the Stone and the Crimson-pearl Flower. Chen greeted the two reverends and asked them to enlighten his benighted understanding with a fuller account of the event. They revealed no heavenly mystery but showed Chen a clear beautiful jade, which had the words "Precious Stone of Spirit" inscribed on one side. Then they went in a stone archway, which had inscriptions of "Illusion Land of Great Void" on the top and a couplet on the side column:

> "Truth becomes fiction when fiction is true;
> The Real becomes unreal where unreal is real."

Chen was about to cross the archway when a sudden clap of thunder woke him up. That same day as Chen took an afternoon stroll with his baby daughter Lotus, they ran into a mangy monk and a lame priest. As soon as the two reverends saw Lotus, they burst into sobs and warned that the baby girl was an ill-fated creature destined to involve her parents in her own misfortune. Failing to persuade Chen to give her away, they left with a verse in puzzling words:

> "Fond man, your pampered child to cherish so-
> That caltrop-glass which shines on melting snow!
> Beware the high feast of the fifteenth day,
> When all in smoke and fire shall pass away!"

The strange pair agreed to meet each other on the northern hill some ninety years later to bring the passionate fairies and amorous spirits back to the Palace of Goddess Disillusionment, and departed for their different terrestrial journeys.

In the Gourd Temple, there was a poor young scholar called Chia Yü-ts'un, who on his way to the capital in search of fame and wealth found himself broken in dire poverty. He lodged in the temple and kept himself alive by working as a copyist. Seeing that it was a waste of genius, Chen assisted Chia fifty taels of silver to go on his journey to the Spring Academic Examination. The ambitious young man thereupon set out for the capital without any delay. Soon it was the festival of the First full Moon. Lotus was carried out by a servant to see the lantern parade and was abducted away by some ruffians. Chen and his wife were nearly driven mad by grief. Then, again, on the fifteenth of the third moon, there was a big fire in the Gourd Temple and Chen's house was also burned into a heap of rubble. They moved to live in a farm in the countryside, but was soon forced by a wild-spread brigandage to seek refuge with his father-in-law. Before long he found that he had little resistance to the joint onslaught of dire poverty and ill health. One day, leaning on a cane along the road, he ran into the lame priest. For the first time he was able to understand the priest's mockery song. Perceiving the true nature of life and human affection he walked into the wilderness with the priest.

In the meantime, Chia Yü-ts'un passed the Imperial Academic Examination and had served as official for years before he was promoted to Magistrate of Ju-chou District. On his arrival he went to pay respect and gratitude to Chen and was informed of Chen's ill-fated lot. He, therefore, promised Lady Chen to search for Chen and Lotus. Nevertheless, one year later, due to jealous slander and unfavourable report Chia was dismissed from his office. After settling his family he set off on an extended tour of celebrated places of scenic interest in the south until he found himself short of cash in the city of Yang-chou. He took the job as a tutor to Black Jade (the incarnation of Crimson-pearl Flower), the only daughter of Lin Ju-hai, the Imperial Salt Commissioner. Upon her mother's death, Black Jade was escorted by her tutor to go to the capital to live in the Mansion of Duke of Glory with her grandmother on her mother's side, while under the assistance of Black Jade's uncle, Chia Yü-ts'un regained a magisterial post in Ying-tien District. There the first case he dealt with was that involved with the abduction of Lotus (now a teenager), sold as a slave girl to Hsüeh P'an (a cousin german of the mansion), who escorted his mother and sister Precious Clasp (incarnation of one of the leading passionate fairies) to visit Aunt Wang, wife of the younger grandson of Duke of Glory. The Mansions of Duke of Glory and that of Duke of Peace were constructed by the imperial command for the two Chia brothers who made great contributions to the empire. Now three generations later, Chia Ching, grandson of Duke of Peace, held a hereditary post as a Third Rank General; Chia Sheh, the elder grandson of Duke of Glory, held a hereditary post as a First Rank General; while Chia Cheng, the younger grandson, was a secretary in the Ministry of Public Works. Though they were not in great power, the Chias were still very wealthy and prestigious elites.

With the arrival of Black Jade, Precious Clasp, and River Mist, the second cousinhood of the Chia clan was heightened; sister-cousins and their maids enjoyed a happy togetherness. Chia Cheng's son Pao-yü—literally Precious Stone, for he was born with a jade inscribed "Precious Stone of Spirit"—was especially delighted, for by nature he had always been fond of female companions.

One afternoon, white visiting the Mansion of Duke of Peace, Pao-yü took a nap in the bedroom of his nephew's wife, Sweetheart, and presently entered the Illusion Land of Great Void. There he was shown the fate of some fifteen young ladies and maid servants of the Chia mansions. It was recorded in verse and emblems on the Scroll of Twelve Beauties of the Capital City. As a teenager, Pao-yü did not perceive the import in those riddles and emblems on the scrolls. Then the Goddess invited him to a feast with a troupe of fairies to sing a set of songs called "the Dream of the Red Chamber," each song an elegy on a beauty or an event. It was, nevertheless, still beyond Pao-yü's ability to comprehend the implications in the songs. Thereupon, after berating Pao-yü for being the most affectionate of all men on earth, and warning him of a more perilous fate because of his potential commitment to a kind of devoted spiritual love to all girls, the Goddess introduced him to her sister for sexual initiation, in order to make him understand the true nature of affection, beauty, and love, and thus to turn his efforts and devotion to the teaching of great sages and the welfare of mankind. Thus Pao-yü met a fairy maiden, a celestial counterpart of Sweetheart who was the incarnation of all beauties, and was instantly struck by the fact that she united in person the charms and grace of both Black Jade and Precious Clasp.

Thereupon Pao-yü had a blissful union with Sweetheart, but was soon dragged away by demons and wild beasts into the Abyss of Illusion to wake up to find it but a dream. As a sexual innocent, Pao-yü again foiled to discern love and other human attachments as a source of delusion and was unable to avoid future involvements in the fate of the major heroines. Thus the didactic dream was followed by a scene of realism in which the forewarned hero coaxed his maid Pervading Fragrance to have sexual intercourse with him to imply the irony of the incorrigibility of man to the delight of sex.

Before long, Pao-yü's eldest sister Cardinal Spring, who had been serving as a Palace-lady-in-waiting, was promoted to Chief Secretary to the Empress and became the Emperor's Concubine. The Mansion of Duke of Glory thus built a temporary imperial residence called "the Garden of Great Wonder" to receive Her Highness for a family reunion. The garden later became the retreat of Pao-yü and his sisters and girl-cousins at the command of Her Highness, who wanted them to enjoy fully the kind of friendship and affectionate warmth denied her in the imperial harem. Indeed, all the passionate fairies and amorous spirits on the scrolls, appendixes, and supplements of the Twelve Beauties of the Capital City were all finally assembled in the garden as their youthful paradise.

As the paradisiac garden was besieged by an ever more realistic world outside, so the innocent adolescents were hunted by an ever more frightening adulthood with all its implied consequences of corruption and unhappiness. Pao-yü, the true host of the garden, naturally wished to liberate all girls from the necessity of marriage with the illusion that the demon of marriage could be exorcised as long as he and his beloved relatives could live for arts in the serenity of the garden. Though he was much obsessed with Black Jade and Precious Clasp, his love and compassion for all the other girls in the garden was tremendously equal and enormous. Nevertheless, the discovery of a pornographic embroidered purse revealed the invasion of lust to virginland; the serpent to the paradise, and the realistic outside world to the idealistic garden. It precipitated the descent of the adolescents from the garden. From then on, unfortunate happenings plagued the household, sorrows and sufferings befell on the fairy inhabitants of the garden, and Pao-yü gradually fell a prey to melancholy, sorrow and suffering.

Thus stunned by deaths of Sweetheart and Bright Cloud and the tragic marriage of Welcome Spring, Pao-yü soon reached his affectionate nadir as well as spiritual zenith, symbolized by the mysterious loss of his jade, his subsequent illness and idiocy. Then Her Highness Cardinal Spring passed away and Pao-yü, in the state of a trance and under the plot of his elders, married Precious Clasp instead of Black Jade, whom he loved most. The event brought Black Jade's death and the sorrow of the household was deepened by Quest Spring's departure for her marriage in the south.

Meanwhile, after his distinguished service as a magistrate for years, Chia Yü-ts'un was promoted to Prefect of the Capital City. One day on his field trip, he came across Chen Shih-yin in a little hut adjoining a deserted temple beside the Ferry of Realization District. Chia failed to talk Chen out of his reclusion; but Chen promised to meet Chia again in a certain ferry. As Chia was crossing the river on the ferry, Chen disappeared into a sudden fire which burned down the temple and the hut into ashes.

Then the Mansions of Duke of Glory and Duke of Peace were impeached for corruption and interference with local administration. Houses were raided and properties were confiscated. Deprived of their hereditary posts and titles, Chia Sheh was exiled to serve in a military fort in the northern frontier and Chia Cheng to serve in the imperial coast guard. The disaster was furthered by the subsequent deaths of Phoenix, Welcome Spring, Grandmother Chia and Lovebird and the abduction of Pretty Jade. The mansions were declining, the garden became deserted to rank weed; and Pao-yü, restored to sensibility and sentiments by Precious Clasp and Pervading Fragrance's attendance, was able to discern the transience and illusion of life through the tragic events, but was dying of grief. Then a mangy monk came to rescue him with the lost inscribed jade. Upon the return of the jade, Pao-yü instantly recovered and subsequently had a dream in which he was guided by the mangy monk to ascend the Bliss Land of Ultimate Truth, where he was received as Divine-stone-in-waiting, and introduced to Black Jade; a goddess now, and most of the deceased female inhabitants of the garden; now fairies. They did not only renounce him but transformed themselves into assaulting demons to be exorcised by the mangy monk who, after instructing him that all the secular affections or human attachments were but demonic illusion, pushed him down to earth. Armed with foreknowledge of the Illusion Land and resigned to the remorseless operations of fate, Pao-yü woke up from the dream a changed person determined to sever all human ties. His insistency to return the inscribed crimson jade to its mysterious sender, the mangy monk, symbolised an act of the deliverance of the self following the attainment of enlightenment. His calm admiration for Compassion Spring and Purple Cuckoo's decisions to turn nuns foreshadowed his renouncement of sensual self and active compassion to release himself from a long obsession with suffering. Yet the solicitude and distress of Precious Clasp and Pervading Fragrance pained him so much that while seeking for personal salvation he resumed his ethical responsibility with commiseration. Then after restoring the Mansions to honour, wealth and power, and having Precious clasp pregnant with a baby, he eventually cut off all human ties. Thus bowing to his homecoming father with a sad smile in a ferry on a snowy night, he walked with the mangy monk and the lame priest into the white and peaceful wildness, leaving a song echoing in the snowy night:

> "The Peak of Greensickness is my residence,
> And the impalpable universe is my wandering place.
> Who is coming with me hither,
> And with whom shall I go there?
> Fathomlessly, mysteriously,
> I'll return to the ultimate Infinity!"

Meanwhile, found guilty in a case of bribery, Prefect Chia Yü-ts'un was deprived of official rank and post. Then on his way home, in a little hut by Ferry of Disillusion, he encountered his former patron Chen Shih-yin; now a reverend Taoist. Reverend Chen came to transport his daughter Lotus who had become Precious Clasp's sister-in-law and died in childbirth back to the Palace of Goddess of Disillusionment. Reverend Chen thus revealed to Chia the secret of the correlation between the terrestrial story of the Garden of Great Wonder and the celestial drama of Precious Stone, Crimson-pearl Flower and all the other amorous spirits. Through the drama of the Mansion of Duke of Glory and his own experience, Chia was finally able to perceive certain true aspects of life, to gain a peace of mind, and to be able to sleep soundly in the little hut by the ferry.

Countless aeons went by, before Taoist Vanitas, passing by the Peak of Greensickness in the Fathomless Crags of Mount Chaos, caught sight of a large stone, on which there was a long inscription. It revealed an account of the affectionate life of the Stone. Perceiving in the story certain truth different from those other romantic stories, Vanitas copied it down and brought it back to look for an editor or a publisher. Contemplating on passion, form, and illusion, Vanitas was able to perceive "void" as the ultimate truth. He therefore changed his name from Vanitas to Monk Amour, and the title of the long inscription from "the Story of the Stone" to "the Tale of Monk Amour." Later on Ts'ao Hsüeh-ch'in worked on it for ten years and subtitled it "The Twelve Beauties of the Capital City."

To regard the novel from the viewpoint of mythopoesis is to discover a structure which recalls Robert Harrison's "quest" progression: a call to quest, an acceptance and descent to the underworld or the time of trials, and a fulfilment of the quest and return in an apotheosized state.[11] Given the novel's complexity, there are several parallel "calls to quest" which occur in both celestial and terrestrial realms. Celestially, it consists of the Sacred Stone's sorrow upon being rejected as a constituent of the Eternal Dome of Heaven, and his acceptance of a life in the terrestrial sphere. The earthly counterpart concerns the collapse of Chen Shih-yin's self-conscious indulgence in domestic bliss, and Chia Yü-ts'un's enthusiastic pursuit of worldly fame and wealth. The acceptance of these calls to quest demands both vertical descent and horizontal extension as the author weaves a journey to disillusionment through a terrestrial life of sense and sentiment for all the descending spirits. Horizontally, the author intrigues for both Chia and Chen a progress to ultimate enlightenment through the fragility and transience of illusory fame, wealth, and happiness. Continuing within Harrison's structure, the "descent" becomes the trials and ordeals undergone by man and immortal alike, while the "fulfilment" of the quest means a return for the descending spirits and a spiritual ascent for the earth-bound mortals. Finally, "enlightenment" is seen as disillusionment with domestic bliss and fame or wealth for Chen and Chia respectively, while it becomes the transcendence of sense and sentiment for the Stone and the affectionate spirits.

On one level *The Dream of the Red Chamber* concerns the Sacred Stone's eventual reconciliation with divinity through an ordeal of affection. The initial rejection of the Stone was caused by his excessive sentimentality, a fact supported by his being discarded at the foot of Peak Greensickness, an

---

[11] Cf. Robert Harrison, "Symbolism of the Cyclical Myth in Endymion," in Vickery, *Myth and Literature*, p. 230.

onomatopoeic pun for "the Peak of Sentimental Sickness." As a consequence, the Stone was granted terrestrial descent to be born in the Mansion of the Duke of Glory. At birth, the miniature inscribed crimson jade in his mouth seems to symbolize his spiritual essence of affection. A group of passionate fairies and amorous spirits were also granted descent to be sisters and cousins to this most affectionate hero, and later to join him in the Garden of Great Wonder to participate in the quest for a life of true affection. Much of the novel is then concerned with providing experience which would enable the Stone to free himself from his obsession with affection, to perceive the illusory nature of the life of the senses and sentiments, and finally to reascend to the spiritual world in an apotheosized state.

Discerning the symbolism of jade in the novel has been a favourite concern of scholars in the past centuries of criticism (and the amount of scholarship is now truly awesome). Wang Kuo-wei, the first Chinese scholar to speculate that jade represents the desires of the self or the will of being, professes that *The Dream of the Red Chamber* is a postulation that the source of the suffering of life is the innate desire immanent in human existence and that the only way to salvation is by self-decisive renouncement of desire.[12] In fact, the elimination of desire and the liberation from attachment have always been the central feature of Chinese religious sects. Furthermore the Empirical School of Confucianism (1644–1912) maintained the necessity of renouncement of affection through the actual experience of involvement. In the Chinese novel, this empiricism inevitably demanded greater realism; detailed description and descriptive narration of everyman's domestic life of sense and sentiment in a vivid psychological context. It is this realistic compulsion which urged Ts'ao to write of personal experience in a striking confessional tone to capture its most intimate reality.

In order to establish the necessity of the renouncement of desire, it is most natural that the Sacred Stone should descend to the Mansion of the Duke of Glory, and Pao-yü to the Garden of Great Wonder, for they are the most sentimental of beings in heaven and on earth respectively. In a sense both descents represent the loss of celestial paradise and the fall of mankind, and reconciliation is possible only through a redemption of real involvement in life. The Illusion Land of the Great Void and its counterpart on earth, the Garden of Great Wonder, are literary devices necessary for integrating the structural application and thematic implication of the novel's empiricism. In the Garden, Pao-yü is just too devoted and too attached to comprehend the intimations of the symbolic scrolls and songs, or the illusory nature of his sexual initiation with

---

[12] Cf. Wang Kuo-wei, "Criticism on the Dream of the Red Chamber," in Yi Li, ed., *Ku-tien wen-hsüeh yen-chiu tzu-liao hui-pian* (Anthology of Materials for the Study of Classic Literature) (Peking: Chung-hua shu-chü, 1963), I, pp. 248–52.

Sweetheart, who is the incarnation of all beauty in the world. Perhaps the "fall" is all too willing, too human; the seduction simply justifies the empirical position that life can only be realized by plunging into the stream of being with full participation, and implies a sexual awakening in the life of the hero ready for the life of true affection in the paradise of beauties. The seduction is the first note in the fateful symphony which will expose the life of sense and sentiment as an abyss of illusion.

With his equal and enormous compassion, his love for one and all, and the idyllic life in the garden, Pao-yü's greatest wish is to liberate his "garden-girls" from the corruption of adulthood and marriage. Naturally this wish is severely undercut by the strong illusion and impermanence of the Garden of Great Wonder. The illusion resides in its analogy and superimposition with the Illusion Land of the Great Void, the dream land of the hero, the dwelling place of the Goddess of Disillusionment, and the final destination of the fairy girls in the garden. The impermanence lies in the steady encroachment of the "real" outside world and the inevitable growth of the garden's inhabitants. Finally the emergence of the pornographic embroidered purse reveals the entry of lust into the virgin paradise, and precipitates the loss of this temporary Eden through the calamities which follow. The sufferings of the hero must be manifold and miserable before he comes to realize the transience of life and indefinite affection; spiritual enlightenment is symbolized by the mysterious loss of his inscribed jade, representative of the desires of the self. The loss of the crimson jade significantly implies in the hero a zenith of transcendence illustrated by the nadir of emotional insensibility and idiotic indifference.

Beset with misfortunes, Pao-yü's evolving renunciation of affection is well-illustrated by scholarly research into the onomatopoeic allusions of the protagonists' names. According to Tu Shih-chieh, "Pao-yü" may be explicated as the "stirring mind," "Black Jade" as "passionate desire and the turmoil of the mind," and finally the most sensible and sensuous "Precious Clasp" as "wisdom or mental tranquillity."[13] In this connection the travesty of the hero's marriage to Precious Clasp instead of Black Jade signifies a timely substitution of wisdom for affection. Furthermore the death of Black Jade heralds the elimination of desire presaged in the loss of the crimson jade. Deprived of the desires of life yet remote from the truth of being, Pao-yü is ironically dying from the emptiness of existence, a notion reinforced by the confiscation of the two mansions and further deaths in the household. At his lowest ebb the hero is rescued by the reappearance of the mysterious monk, this time bearing the lost crimson jade. In a dream Pao-yü is transported to the Bliss Land of Ultimate Truth (significantly subtitled the Disillusion Land of Affection), where he is

---

[13] Cf. Tu Shih-chieh, *Hung-lou meng pei-chin tao-yü shih-k'ao* (Taichung: Author, 1971), pp. 363–65.

informed of his mythic origin as the Sacred Stone, reacquainted with the significance of the scrolls, and escorted to the company of the deceased Black Jade and the elevated spirits of the other fairy girls from the ruined Garden of Great Wonder. Such revelations compel the hero's realization of the illusory nature of life and the afflictive nature of affectionate attachment. Hence, his enlightenment was signified by his desire to return the crimson jade to its sender, and his determination to sever all human ties and thus release himself from the obsessions of suffering. Thus, after restoring the mansions to honour, wealth and power, and confirming Precious Clasp's pregnancy, he retires into the white wilderness with a rejoicing song of homecoming.

The sub-plot concerning the parallel quest of Chia Yü-ts'un was similarly resolved, as this incessant seeker of fame and wealth was once more deprived of official rank and honour and came across his former patron, Chen Shih-yin, who is in the apotheosized state of a divine ferryman. Here, Chia is able to discern the epiphanic meaning of his encounters with Chen, the lifelong recluse and ferryman of true being. He perceives the truth of life in the myth of the Sacred Stone and the drama of the Garden of Great Wonder, recited to him by the divine ferryman. At last, he is able to sleep soundly in the little hut by the ferry.

The episodes of the mansion's restoration as the lingering temptation of fame or wealth and Precious Clasp's pregnancy as the ultimate affectionate seduction may be seen as literary devices which reinforce the tension of the tragic pathos of the hero's renunciation. They also imply the philosophical application that the approach to Nirvana is the ultimate discarding of whatever means by which it was brought forth. Thus either as an image of sensuous beauty and marital devotion, or as a metaphor of wisdom (a means to enlightenment), Precious Clasp's fate of desertion is necessary and inevitable as an affirmation of the hero's triumph over worldly attachments. It is in this connection that the Garden of Great Wonder may be regarded as thematically equal (in an operative sense) to T'ang Hsien-tsu's dream world inside the pillow, and similarly that the conversion of Vanitas to Monk Amour as an expression of the novel's theme is identical to that of Wu Ch'eng-en's Heart Sutra, which professes that only in the quietude of mind and heart is Nirvana accessible.

Certainly the epilogue of Chia's enlightenment as an allegory of disenchantment with the ephemerality of fame and wealth is both a contrast and a complement to the mythic drama of the Sacred Stone as an allegory of disillusionment with affectionate attachment. The testimony of the seeker and the recluse in the finale concerns neither their personal vicissitudes nor those of the Mansion or the Garden; rather the subject must be the flux and reflux of life as a whole. The epilogue of bitter realization and renouncement of being is paradoxically the prologue of the quest for non-being and reconciliation with the infinite.

The Dream of the Red Chamber reflects a particularly Buddhist view of concepts underlying the cyclic myth. One would expect that a life of harmony with cyclical nature would be the aim of life within the Garden of Great Wonder. But of course the Garden is established not so much as a metaphor of natural harmony but as a dream world wherein Pao-yü may pursue his ideal of unchanging self-conscious affection far from the everchanging mundane world. Inasmuch as this precept ignores the cyclical impermanence of reality it is doomed to tragic failure. The transcendental solution depends upon the revelation that because reality is transient and ephemeral, human concerns must reflect the Tao which underlies that reality. Nirvana, or the unity of the self with the cosmos, must be consequent to the revelation of life's ephemerality. Ts'ao Hsüeh-ch'in's obvious fondness for his imaginative creation, the Garden of Great Wonder, implies the difficulties of the way to transcendence. Indeed, as C.T. Hsia interprets the author of *The Dream of the Red Chamber* is "a tragic artist caught between the nostalgia for, and the tormented determination to seek liberation from, the world of red dust."[14]

4. Summary: the Cyclic Schema in Chinese Poetry, Drama, and the Novel

Within the range of Chinese literature selected, whether as poetic contemplation, dramatic manipulation, or novelistic philosophical postulation, aspects of the "cyclic" mentality continue to appear in an integrative fashion throughout the literature of China. Concerning the evolution of the cyclic myth, a general progression is evident: from the earliest mythical envisagement through ritual actualization to philosophical rationalization, and finally through various literary applications or expressions. Given the complexity of Chinese literary culture and the vagaries of historical scholarship the above evolutionary pattern should not be strictly interpreted in either a causal or limiting temporal sense; rather the pattern corresponds generally to the trend of increasing intellectuality of Chinese culture.

The proposition that the cyclic myth has been one of the prime integrating factors throughout Chinese literature is reinforced by the continual reappearance of the following archetypal images:

a. Archetypal Images

(1) **The image of mountain** as the cosmic mountain, Mount K'un Lun, the geographical link between divine and mundane consciousness. Such mountain images as the Southern Hill, Mount-to-heaven, Mount Flower-and-fruit, Mount Holy Terrace-to-the-heart, Mount Five-elements, the Holy Mountain of Buddha,

---

[14] Hsia C. T., *The Classic Chinese Novel*, p. 297.

and Mount Chaos all connote the symbolism of Mount K'un Lun as the cosmic mountain and the passageway to Heaven. As such it is the destination of the initiating protagonist and the questing hero who would obtain reconciliation with divinity, eternity, or the cosmos. When the "mountain-quest" is undertaken in bad faith, punishment is typically an inversion of the above sequence: estrangement or banishment. Thus, for example, Monkey is banished from Heaven and pressed under the cosmic Mount Five-elements.

(2) **The image of water** as the cosmic water, the Sinking Water or the Lake of Purity. Such water images as the Sinking Sand bounding Earth from Heaven, the Sacred Water surrounding the Western Paradise of Buddha, the White Water, the Black Water, the Red Water, the Lo River, the River-to-heaven, the Eastern Sea, and the Transmutation-lake-for-dragons all connote either the Sinking Water as the sacred obstruction to the realm of immortality, or the Lake of Purity as the Water of Life for the divine thirst or as the abode of the Supernal Mother Goddess. The attributions of death, life, purification, and immortality of this cosmic water are best illustrated by Tripitaka's astral projection in the bottomless boat upon his crossing the Sacred Water surrounding the Holy Mountain of Buddha.

(3) **The image of garden** as the Garden of the Tree of Life, the Hanging Garden. Such images as the House of Spring, the Garden of the Sacred Peach, Mount Flower-and-fruit, the cultivated garden of orchids, the self-contained farm on the Southern Hill, the garden of *The Peony Pavilion*, the garden of the Ts'ui, and the Garden of Great Wonder all connote the original Hanging Garden with its attributes of natural beauty, its sacred fruits and divine boughs of the various Trees of Life, its nubile goddesses and beauties, life-sustaining foods, pastoral happiness, and paradisal bliss. Again, the Hanging Garden is the uppermost circle of Mount K'un Lun and is the home of the Tree of Life and the source of various divine ambrosias and magical nectars. Thus Monkey's role as hero of a cosmic quest is best explicated by the paradox of his being both the superintendent of and the intruder in the Garden of the Sacred Peach.

(4) **The image of spring or autumn** as the symbol of cosmic transformation. Images of seasons all connote the various phases of bio-metaphysical transformation consonant with the Myth of Divine Administration. Spring and autumn with their strikingly visible transitions are specified as symbols of the ephemerality of life and the transience of being. Thus the temporal structure of *The Peony Pavilion* determines its thematic structure, and the inevitable fall of the Garden of Great Wonder as the virgin land of ideal paradise is similarly foreshadowed. As the hostesses of the Garden of Great Wonder are termed the maids of spring; Cardinal Spring, Welcome Spring, Explore Spring, and Farewell Spring, their nomenclature prefigures the inevitable seasonal transition. In addition the enormous celestial army sent to arrest Monkey is composed of

the cyclic powers of time and space (including the four Gods of Season), and the unavoidable transition is again suggested.

(5) **The image of the heroine** as a mate or a patroness. Images of primary goddesses such as Nü Wa, Hsi Wang Mu, river nymph, the Goddess of Flowers, the Bodhisattva of Mercy, and the Goddess of Illusion; or leading heroines such as Fair Bride, Miss Ts'ui, Sweetheart, and Precious Clasp, all connote the Supernal Mother Goddess who is also the Goddess of Marriage and is usually in charge of the sacred fruits and waters, and the elixir of life. This feminine image is hence the mate of the initiating protagonist or the patroness of the questing hero. Her function is essentially that of completion and is borne out by the inevitable pairing of Yin and Yang as the masculine and feminine principles in the earliest development of *The Book of Changes*.

(6) **The image of the recluse** as the symbol of the diviner or the immortal. Images of recluses such as Wu Hsien, P'eng Hsien, Ling Fen, Taoist Lü, the Abbot of the Golden Mountain, Zen Master Crow-nest, the mangy monk, the lame Taoist, and Taoist Chen all connote the diviner who is the seer of ultimate truth and is hence the patron or guide of the hero in his time of trials. He is often the ferryman in the questing hero's final ascent to reconciliation.

(7) **The images of dream and death, and the dream-world** as the symbol of descent and the time of trial. Images of dream-worlds such as T'ao Ch'ien's vision of the mundane world, the limbo of Fair Bride, the dream-world inside Lu Sheng's pillow, Tripitaka's illusory land of demons, and even Pao-yü's Garden of Great Wonder all connote a life of extraordinary sense and sentiment, which is at the same time a trial that will provoke a bitter realization of the transient and illusory nature of being. Similarly such images as Fair Bride's death, Lu Sheng's dream, Monkey's death, Emperor T'ai-tsung's death, Golden Cicada's exile, Monkey's fall into the Crucible of the Eight Trigrams or his imprisonment under Mount Five Elements, Pao-yü's dreams of the Illusion Land of Great Void, or his and other amorous spirits' exile all connote a descent into trial and ordeal.

(8) **The image of awakening or resurrection** as the symbol of the return of the questing hero in an apotheosized state. Such images of awakening and resurrection as Fair Bride's resurrection, Lu Sheng's awakening, Monkey's awakening, Emperor T'ai-tsung's resurrection, and Pao-yü's awakening all connote the symbolic return of the hero in an apotheosized state. Frequently such returning heroes possess a divine ability or spiritual transcendence which enables them to achieve the final ascent to enlightenment. Thus Tripitaka's return on a puff of fragrant wind and Pao-yü's final recovery (or awakening) from illness and idiocy are allegorical in this respect.

(9) **The image of conversion or canonization** as the symbol of the achieved ascent to reconciliation. Such images as Ch'ü Yüan's immortalization in universal roaming, T'ao Ch'ien's spiritual transcendence, Fair Bride's final

resurrection and paternal reconciliation, Lu Sheng's conversion into the immortal island, the canonization of Tripitaka and fellow pilgrims, and Pao-yü's conversion to the recluse of Mount Chaos all connote the achievement of final reconciliation with divinity. This ultimate apotheosis is best delineated by Tripitaka's canonization as manifold reconcilement; other examples include the exiled god with divinity, the pilgrim with Buddha, and the individual with the cosmos.

The above recurring images essentially depict a mentality which looks inward and outward at the same time. The severely nostalgic vision of alienation and exile looks inward into a heterogeneous world of bio-cosmic estrangement; yet by assigning equal ontological reality to both microcosm and macrocosm and by affirming their mutual application, it looks outward to a world of universal wholeness. Such a mentality immediately transcends the boundaries of literature and philosophy, of myth and ritual, and presents the primitive psyche attempting to satisfy a nostalgia for temporal duration and create a harmony between man and the cosmos. If we consider the psyche which has created and recreated the cyclic myth through the forms discussed, certain symbolic phases become evident. Here we are not concerned with specific images from literature or concepts from philosophy so much as with a progressive scheme or pattern, symbolic in a general sense, which the Chinese mind has adopted and adapted in its search for reconciliation. Thus the cyclic envisagement consists of several 'symbolic' phases.

b. Symbolic Phases

(1) **The call to the quest** is usually evoked by the tristia, a cosmic sorrow, arising from a realization of the ephemerality of life, and the illusions of the world of sense and sentiment, or from the temporal awareness of the finitude of man in his cosmic estrangement. The tristia depicts a linear consciousness, an individual alienation, and a nostalgic desire for cyclic harmony. It therefore arouses a desire for permanence within the everchanging cosmos and unity with a universal oneness which in turn motivates the cosmic quest.

(2) **The quest** delineates a cosmic itineraria, a temporal and spatial journey or progress, from the "profane time" of the terrestrial world to the "great time" of the demonic or underworld and finally to the "no time" of the celestial world. The progress may elaborate on such typical quest patterns as the search for the waters, fruits, and elixirs of life which guarantee cosmic perpetuity for the partaker; an incessant pursuit of a goddess as mate to restore a balanced harmony of opposites; or a panoramic circuit of the universe, the conclusion of which provokes an apotheosis. The journey itself may elaborate either a descent to the land of ordeals for purgation by trial, or an ascent to the hub of the universe and the abode of divinity for a sacred reconciliation.

The quest for a goddess elaborates upon an incessant pursuit of marital union or divine reconciliation with a legendary beauty, a mysterious nymph, or fairy goddess. All are representative of the Supernal Mother Goddess who is the patroness of marriage and of the fruits, waters, and elixirs of life. The token of love and proposal is often a fragrant blossom, a sacred stalk, or a divine branch of jasper leaf. The symbolism of vegetable adornment and nutriment expresses a contagious and reflective insight into the interpenetration of human microcosm and natural macrocosm. It often concentrates on the "mana" of food through the fruits and waters of immortality, or the sacred power of the golden bough (as a fragrant blossom, magic stalk, or a divine leaf or branch).[1]

(3) **The return of the hero** depicts a fulfilment of the quest in terms of a resurrection from death, an awakening from dreams, a renunciation of the mundane world, or a transit to the immortal realm. The return is made in an apotheosized state and in turn justifies the hero's final reconciliation with the cosmos.

(4) **The reconciliation** expresses the actualization of the original wish-fulfilment. It may assume such forms as the reconciliation of a fallen god with the supremacy of ultimate divinity, of man with nature as a harmonious unity, of the pilgrim with Buddha, or the self with existence as a spontaneous oneness. Reconciliation may be indicated by conversion, canonization, or transcendence; its actualization elucidates a coherent application of the mutual relationship between man and god or universe. This integration is central to the homogeneity of Chinese culture.

As a whole the cyclic pattern delineates the psyche of estrangement and reintegration between the fallen status and that of restoration, the lost paradise and that to be regained, and the severed relationship and that of reconciliation. It is designed to soothe a linear consciousness and integrate the heterogeneous reality of life and being, and is evident in all of the literary works selected above.

If myth is regarded as the lowest common denominator of cultural expression, or as a narrative pattern informed by the elemental drives and anxieties of the collective psyche, then our elaboration of the Chinese cyclic myth bears many similarities to Robert Harrison's outline of the Western cyclic myth offered below:

> "The cyclical myth may be described as a spiralling motion in which the hero experiences a decent (kathodos) and an ascent (anodos), emerging not at the point of outset, but at a higher

---

[1] Although the Chinese myth of the golden bough is unsubstantiated, apparent fragments are to be found in the ancient classics. Thus the proverbial phrase that a beauty's genesis is of "the golden bough and the jasper leaf " must be more than figurative speech.

level. The basic movements in this eccentric circle are: The Call to the Quest (for it is a quest, though not a Neo-Platonic one), Acceptance and Descent into the Underworld (time of trials), Fulfilment of the Quest, and Return, often apotheosized by some sort of sacred marriage."[2]

Our examination of the progression of myth, ritual, philosophy, and finally literature has posited a homogeneous cyclic schema present in every sphere of human expression in Chinese culture. To the Chinese this is the concept of **"T'ien Jen Ho I,"** the unity of man and the cosmos.

---

[2] Robert Harrison, "Symbolism of the Cyclical Myth in Endymion," in Vickery, *Myth and Literature: Contemporary Theory and Practice*, p. 230.

III. ESSAYS ON WESTERN TEMPORAL SCHEMATA IN MYTH, RITUAL, PHILOSOPHY, POETRY, DRAMA, AND THE NOVEL

A. THE SURVIVAL OF CYCLIC MYTH IN WESTERN LITERATURE

1. Historical Background: Linear Eschatology Sustained Through Cyclic Symbolism

Obviously no account of the survival of the cyclic myth in Western culture would be complete without some mention of the forces which brought about its radical shift. To begin, the extent to which the developing Hebrew faith conceived of man's relationship to divinity rather than his identification with that divinity is the earliest measure of the "finite" temporal sense. In the centuries following the decline of Sumerian civilization, continual religious persecution resulted in the formation of a tribal religion unique in its insularity. Jeremiah and Ezekiel are credited with advancing the idea that all religions except one are false, and that the Lord Yahweh punishes idolatry. Meanwhile destruction of the Temple ensured that Judaism would become non-sacrificial with regard to ritual, a marked departure from the practices of contemporary and competing faiths. In time a theology evolved in which man had been created not to enjoy divinity but to know, honour, and serve it. Once this existential focus was codified in the biblical trials of Job and the sacrifices of Abraham, reconciliation with divinity, so common in Chinese theology, lost its immediate possibility and became more future-oriented. The consequent mythology developed away from the notion of periodic regeneration and toward a progressive temporally-oriented worldview of a creation; once and for all, at the beginning of time, a subsequent fall from grace, and a continuing restoration. The world no longer was to be experienced as divinity operative through cosmic rhythms; rather it became a field of cosmic conflict between two powers, one light and the other dark. Finally, as Joseph Needham indicates in his brilliant *Time and Eastern Man*, the consequences of Hebraic Messianism included a radical view of time and history.

> "The Hebrews were the first Westerners to give a value to time, the first to see a theophany, an epiphany, in time's record of events. For Christian thought the whole of history was structured around a centre, a temporal mid-point, the historicity of the life of Christ, and extended from the Creation through the covenant of Abraham to the second coming of Christ, the messianic millennium and the end of the world."[1]

The Messianism of the Hebraic, and consequently the Christian faith, assumes on a higher plane the eschatological role of the king as representing

---

[1] Needham, *Time and Eastern Man*, p. 45.

divinity on earth. In accordance with archaic belief the king's chief mission was the periodical regeneration of all nature. The trials of the Hebrew prophets suggest what is manifest in the life of Jesus Christ—that victory, according to the ancient scenarios, was always finally the king's. The crucial difference is that now this victory occurs not annually but is projected into a Messianic future. Moreover, through the terrible visions of the prophets and the promise of future victory, historical catastrophes were construed as Yahweh's wrath and acquired meaning as revelations or concrete expressions of the same single divine will. Historical facts thus become "situations" of man in respect to God and history itself is seen as the epiphany of God. Such a stern vision was naturally difficult for the elite to inculcate in times of prosperity, and the history of Hebraism is marked by many returns to such Paleo-Oriental divinities as Baal and Astarte in the absence of calamity.

As this essentially linear view of time and history was inherited by Christianity and firmly codified by St. Augustine, Western temporal attitudes became fixed in the linear pattern stressed by Henri-Charles Puech:

> "A straight line traces the course of humanity from initial Fall to final Redemption. And the meaning of this history is unique, because the Incarnation is a unique fact. Indeed as Chapter 9 of the Epistle to the Hebrews and I Peter 3:18 emphasize, Christ died for our sins once only, once for all; it is not an event subject to repetition, which can be reproduced several times. The development of history is thus governed and oriented by a unique fact, a fact that stands entirely alone. Consequently the destiny of all mankind, together with the individual destiny of each one of us, are both likewise played out once, once for all, in a concrete and irreplaceable time which is that of history and life."[2]

Of course the original myth of the eternal return could not be utterly replaced by the linear approach of Judaeo-Christian orthodoxy. The mnemonic power of primal archetypes and the immediate evidence of natural cycles to an agricultural society ensured the survival of the cyclic myth. In fact as Mircea Eliade so ably demonstrates in his *Cosmos and History*, cyclic speculation still occupied a prominent role for primitives and intellectuals alike well into the seventeenth century.[3] Even within Christianity cyclic naturalism thrives in cosmological symbolism which bears remarkable similarities to the Oriental structures previously mentioned.

---

[2] Cf. Henri-charles Puech, cited by Eliade, *Cosmos and History*, p. 143.
[3] Cf. Eliade, *Cosmos and History*, pp. 141–47.

According to the archetypal symbolism of the centre, Paradise or Eden is situated at the centre of the cosmos; it was on this site that Adam was created, and the same spot where the Cross of Christ was erected. This sacred mountain, where Heaven and Earth meet, is the Axis Mundi where the Tree of Knowledge once stood. In the myth of Eden the Tree stands in the centre of the Garden, at the source of the four rivers which metaphorically "water the garden." Eden is not only the mirror of Heaven above; it is also a reflection of Christ, wherein all the events of man's redemption are seen in reverse. Over against the Tree of Knowledge, from which comes death, is the Tree of the Cross, from which came eternal life. One is reminded of the myth which identifies the wood of the cross with a branch or splinter taken from the Tree of Eden. In this manner Christianity weaves a web of symbolic correspondences reminiscent of the abstractionist tendencies of Ming T'ang, the Yin-yang school, and especially the myth of Mount K'un Lun in Chinese thought.

The most significant survival of the cyclic myth in Christianity concerns the cycle of the Christian liturgical year. This religious calendar commemorates, in the space of a year, all the cosmogonic phases observed by archaic man, yet combines these phases within the Christian framework of the birth, death, and resurrection of the Son of God. In a sense the sacred year ceaselessly repeats the Creation; man becomes contemporary with cosmogony as ritual projects him into the "great time" of the beginning. With this cyclic liturgy in mind, Alan Watts suggests a scheme of pagan correspondences:

> "The seasons of the year are themselves transformed from the pagan Spring, Summer, Autumn, and Winter to the Christian Advent, Christmas, Epiphany, Lent, Passiontide, Easter, and Pentecost. However, because the sun itself is seen as a type of Christ, the Sun of Justice, the Christian Year is rather significantly integrated with the cycle of the sun. The Christian Year begins four weeks before Christmas, which coincides approximately with the Winter Solstice— the time when, in the Northern Hemisphere, the sun is at its lowest meridian and is about to begin once more its upward journey to the midheaven. Anciently, this time was sometimes known as the Birth of the Sun, being as it were, the midnight of the year, from which point the sun begins to rise. According to tradition, then, Christ was born at midnight at the Winter Solstice."[4]

This pattern of correspondence could be extended into greater detail. For example the Pentecost, fifty days after the Passover, was originally the Jewish

---

[4] Allan Wilson Watts, *Myth and Ritual in Christianity* (London: Thames and Hudson, 1954), p. 87.

Feast of the Weeks, the celebration of fruition and harvest. The Passover also celebrates Jewish deliverance from Egypt; with the harvest imagery, Christ is often referred to as the "first fruit" of the New Creation. Thus the first to rise from the dead rises in the very season when the buried grain first rises to fruition. Generally, in the cycle of the Christian year the rites of the Incarnation are governed by the solar calendar, since they are connected with the renewal of the solar cycle and fall upon fixed dates. On the other hand the rites of Atonement, Death, Resurrection, and Ascension are governed by the lunar calendar; the images of Death and Resurrection find their natural counterpart in the waning and waxing of the moon.

If the cyclic myths of the Christian world and China can be said to have spatial form, then the figures of the circle and the spiral would be their images. What is most crucial is the relative emphasis on periodic regeneration in each schema: the very heart of the Chinese cyclic myth is the ever-recurring natural cycle whose principle is the Tao which provides liberation from the anguished temporality of linear consciousness; at the base of the Christian format we find the paradigmatic life of Christ, cyclic through birth and resurrection, yet not periodic because of the historicity of his life. To repeat Puech, the birth and death of Christ are not subject to repetition, they are historical events. Thus despite the cyclic symbolism of the liturgical year and the manifold efforts of both laymen and clergy to perceive natural correspondences, Christianity remains essentially a faith of linear eschatology.

2. The Survival of the Cyclic Myth in Natural Symbolism

As we turn to study the Western literary view of the cyclic myth, it must be pointed out that although the ideal of Christianity is a linear eschatology, in practice the cyclic elements receive prominence in relation to the orthodoxy of the practitioner. Thus the thought of Augustine, a member of the orthodox elite, is far more "linear" in its intentions than that of any of the countless heterodox laity of Catholicism. One would be more correct in maintaining that the Middle Ages were dominated by the eschatological conception of the end of the world, complemented by the theories of cyclic undulation which explained a periodic return of events. The survival of archaic rites of periodic regeneration no doubt provided a sense of certainty to moderate the unrelieved misery of the era. Recalling that Christian dogma maintained nothing of value was possible in the sublunary world except steadfast virtue which would eventually lead to eternal bliss, it is not surprising that vegetative rites and superstitions should survive in lore and practice.

By the fourth century A.D. the Hellenistic melting pot of barbarian myths and esoteric religions had been given "philosophical sanction" by the efforts of the Neo-Platonic influence, entailing a separation of historical event and metaphysical element in favour of the latter. Popularly a vulgar spiritualism replaced the historicity of Christ and engendered, in Erich Auerbach's words, "a rather dismal sort of erudition; elements of astrology, mystical doctrine, Neo-Platonism, strangely distorted in the vulgar mind, were summoned up in support of this reinterpretation of events, and an abstruse art of allegorical exegesis was born."[5] Despite this climate of allegorical excess, where every object and event of historical Christianity could be endowed with a "meaning" through importation of natural and mythological symbolism, the literal reality of Christ's life and the transcendent meaning of his love survived. Toward the end of the first millennium dogmatic allegory had been replaced by a spiritualization of the mundane world, from its greatest political developments to its least consequential daily affairs. Medieval man had discovered his intensely personal role in the drama of salvation and an individual fate that was decisive for all eternity. Paradoxically the development of allegory and mimesis, the latter superseding the former, ensured the survival of cyclic symbolism; reality could be presented as the operative sphere of the triune divinity wherein the archetypes of natural cycle and mythological quest were subsumed. Thus, although Dante was the first master of mimesis of secular Christian reality, he was also an encyclopedic symbolist in the exegetical tradition.

3. Dante's *Divine Comedy*: A Spiral Transcendence to the Centre of Circles

While both mimetic and allegorical attitudes in *The Divina Commedia* are indicated by Dante Alighieri (1265–1321) in a letter to Can Grande,[6] and further developed as the "fourfold method" of symbolic interpretation, the poet does not mention that this is also the story of the pilgrim's own growth and development, his quest for enlightenment. Indeed the progress of the narrator parallels the cyclic myth of the hero in its four major phases: the tristia— wherein the poet discovers his alienation and spiritual estrangement in the "dark wood"; the itineraria—wherein he progresses through both a descent to the land of ordeals and an ascent to the hub of the universe; the return of the hero— wherein Dante "returns" to wholeness, fulfilling the sub-quest for the goddess (Beatrice) and achieving apotheosis through an intellectual comprehension of God; and finally, reconciliation—wherein the poet-hero achieves the state of ecstasy necessary for his intuitive apprehension of God. In the course of the

---

[5] Erich Auerbach, *Dante: Poet of the Secular World*, trans. Ralph Mannhaim (Chicago and London: The University of Chicago Press, 1961), p. 18.
[6] "The meaning of this work is not simple but rather may be called polysemous, that is to say, having more meaning than one, for it is one meaning that we get through the letter, and another which we get through the thing the letter signifies; the first is called literal but the second allegorical or mystic." Cited by Thomas Bergin, *Dante's Divine Comedy* (Englewood Cliffs: Prentice-Hall Inc., 1971), p. 78.

journey the soul descends into its own depths knowing that these depths are also those of the created world. Dante's mythopoesis is most clearly shown in his various encounters with the monstrous apparitions which bar his path and guard the way to the Tree of Life (or Knowledge). Encounters with the lion, leopard, and she-wolf, and Minos, Cerberus, Pluto, the Gorgon with her Furies, the Minotaur, Geryon, and the figure of Lucifer himself link the story of Dante's descent into Hell not only with the death and resurrection of Christ and his victory over the Devil, but also with the myths of the great classical heroes—Hercules, Theseus, Perseus, and the myth of the rape and rescue of Proserpina. Such mythologizing reinforces the quest aspect of *The Commedia*, and establishes structural use of the cyclic myth as a governing pattern.

In his monumental analysis, *Dante: Poet of the Secular World*, Erich Auerbach discerns *The Commedia*'s structure to be three great interwoven systems; physical, ethical, and histori-political.[7] The physical system consists of the Ptolemaic order of the universe adapted to Christian dogma by Christian Aristotelianism, and accounts for the geographical outline of reality and the extension of Being from the "Primum Mobile" through all created beings to man who occupies a special position through his exercise of free will. The second or ethical system generally follows Thomist ethics and is responsible for the complex hierarchies of Hell and Purgatory. For example, the terraces of Purgatory are subdivided according to the nature of the love that must be rectified in the individual penitent. Finally within the histori-political system are contained the interpenetrating histories of the creation and fall of man and the Roman Empire, and the world-renewal myths of the Near East alluded to previously. Here we are most concerned with the "physical" system of Dante's world in its cyclic aspects, best demonstrated by the diagrams offered by Thomas Bergin.[8]

---

[7] Cf. Auerbach, *Dante: Poet of the Secular World*, pp. 101-33.

[8] Diagrams of Dante's conceptions of the Cosmos and Hell are reproduced from the facing pages of *Dante's Divine Comedy* by Thomas Bergin, Copyright© 1971 by Prentice-Hall Inc., reprinted by permission of Prentice-Hall Inc.

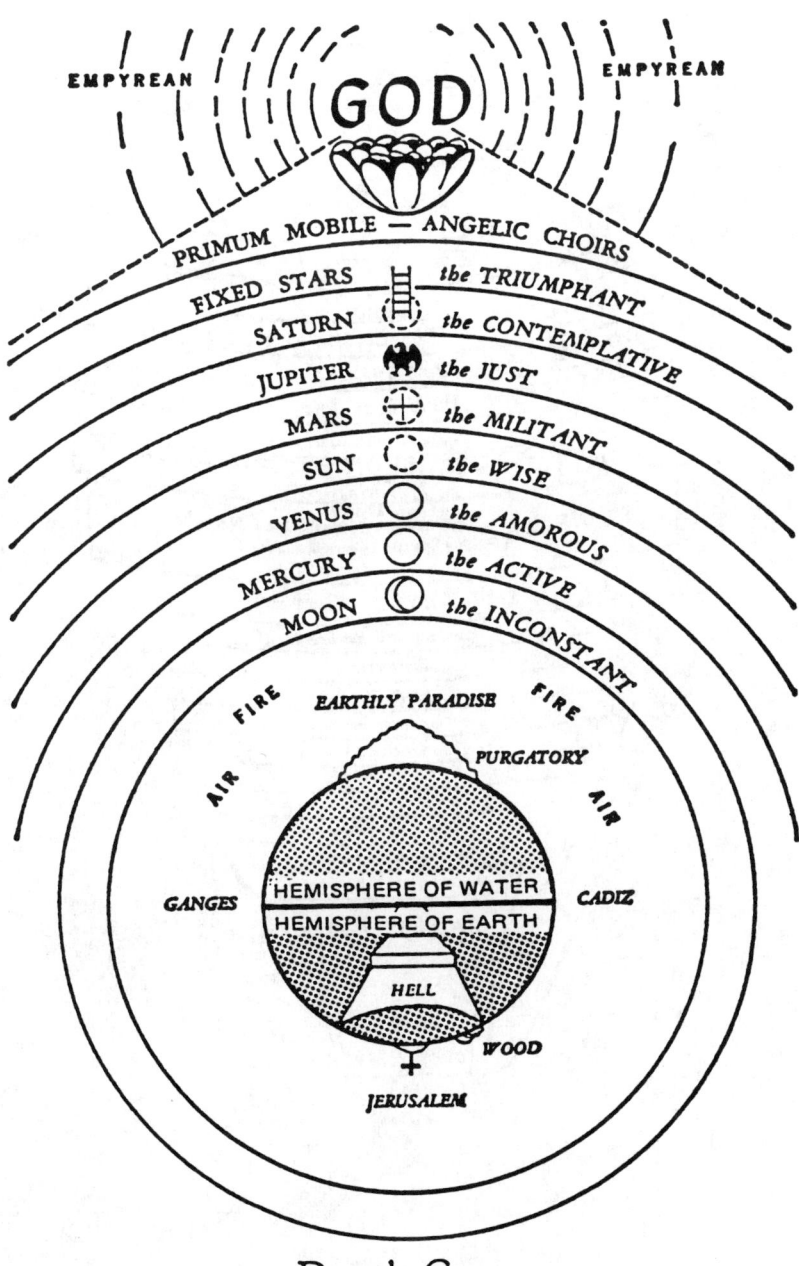

Dante's Cosmos

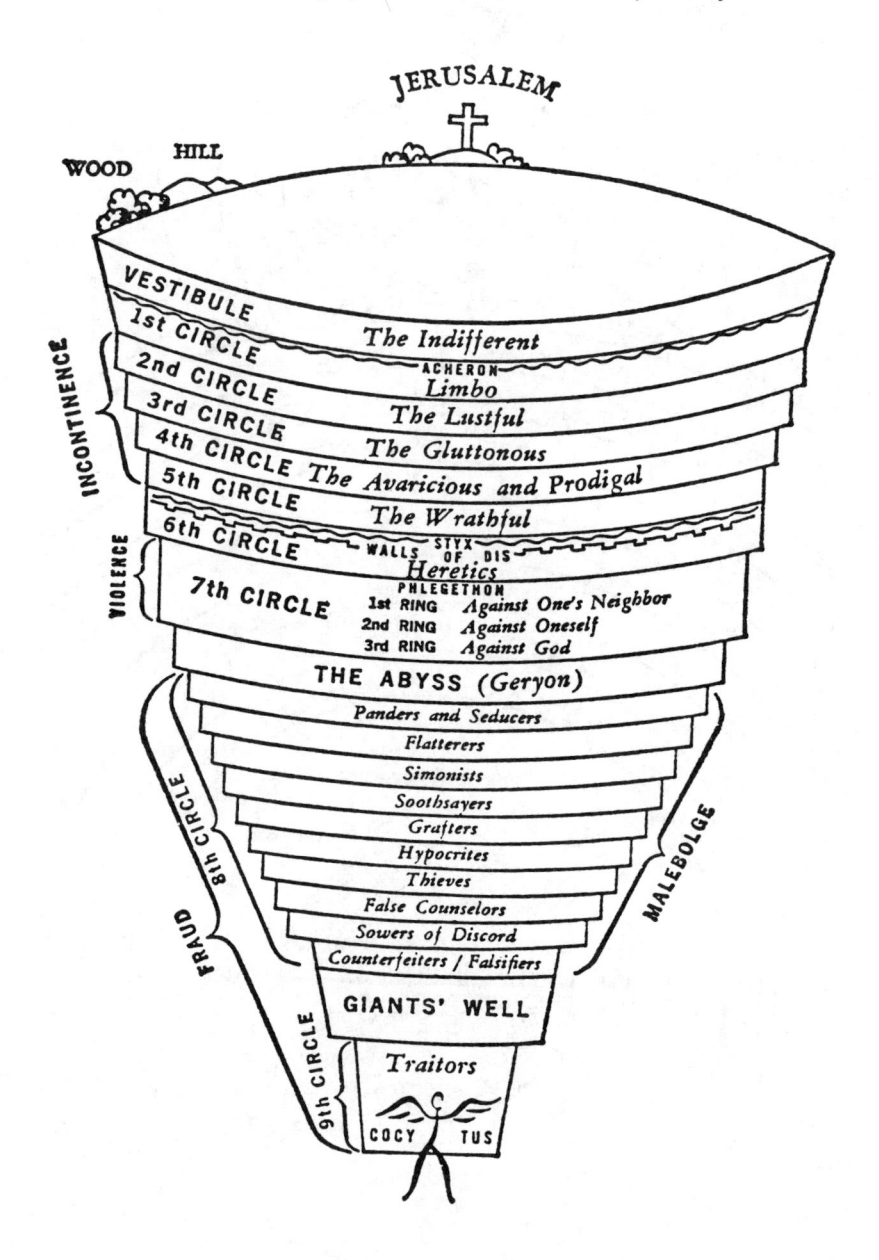

134 A Comparative Study of Chinese and Western Cyclic Myths

JERUSALEM

WOOD HILL

VESTIBULE
1st CIRCLE — The Indifferent
— ACHERON —
2nd CIRCLE — Limbo
3rd CIRCLE — The Lustful
4th CIRCLE — The Gluttonous
5th CIRCLE — The Avaricious and Prodigal
— The Wrathful
6th CIRCLE — WALLS OF DIS — STYX
Heretics
— PHLEGETHON —
7th CIRCLE
   1st RING Against One's Neighbor
   2nd RING Against Oneself
   3rd RING Against God

THE ABYSS (Geryon)

Panders and Seducers
Flatterers
Simonists
Soothsayers
Grafters
Hypocrites
Thieves
False Counselors
Sowers of Discord
Counterfeiters / Falsifiers

8th CIRCLE — FRAUD — MALEBOLGE

GIANTS' WELL

9th CIRCLE
Traitors
COCYTUS

INCONTINENCE
VIOLENCE

Hell

According to these static representations, the spatial form of Dante's world is the circle, but dynamic progress through these circles takes the form of a spiral. In the "Inferno" Dante moves in a narrowing leftward spiral to the centre of the earth-sphere where he overcomes Lucifer. From this negative centre the direction of the spiral is reversed, and in the "Purgatorio" it ascends. Again it is an inward progression to a positive centre represented by the Tree reaching up to Heaven, or the fountain flowing from divine will. In the "Paradiso" the movement is apparently outwards until the sight of God as an infinitesimal point of light in Canto XXVIII; then everything is turned inside out and the heavens appear as wheels turning round the luminous point as centre. Dante no doubt conceived the spiral to represent the relationship between the centre and the circle: through the cyclic interplay of the Holy Trinity, God's love is reflected in the circular motion of celestial bodies, while man's progress to the centre reproduces this pattern in reverse.

Dante was meticulous in locating the action of *The Commedia* in real time, again in a cyclic fashion. The pilgrim's journey from Hell to the Empyrean takes seven days—Easter week, the most important in the Christian calendar. He enters the Gate of Hell on Good Friday, leaves Hell on Easter Sunday, and the journey through Purgatory and Paradise ends on Easter Thursday to complete the circle. In another sense the poet's quest may be said to follow the seasonal cycle. Beginning in the dark wood, the dying world of autumn, he descends to encounter Satan in the icy cold of winter; spring, the ascent phase, is represented in his reunion with Beatrice, and summer with its light and warmth is seen in the poet's Empyrean enlightenment. In this manner Dante defines his cosmology within the natural cycle, a process which is unifying in structure and universalizing in theme.

Apart from these structural uses of the cyclic myth, Dante further grounds his poem thematically in the allegorical traditions of natural symbolism, most notably in his use of solar, pastoral, and lunar or water imagery. In encyclopedic detail Helen Flanders has traced how the syncretism of the first centuries after Christ resulted in a solar pantheism, a trinity of life, light, and heat which became the triune Sun of the "Paradiso."[9] Whether or not Dante's use of solar imagery is a consequence of the assimilation of Mithraism by early Christianity, there can be no doubt of his emphasis on the cyclic aspects of the sun as the image of the rebirth of the Son. In the second canto of the "Purgatorio" the sun rises from the ocean marking the dawning of Easter Day. The sun is of course Christ, the Son, risen in Dante's heart and lighting his way. From this point on, Virgil is more of a companion; the real guide is the light of the sun and its movement around the earth.

---

[9] Cf. H. Flanders Dunbar, *Symbolism in Medieval Thought and Its Consummation in the Divine Comedy* (New York: Russel & Russel, 1961), pp. 105–262.

The vegetative cycle finds a unique application in the Christian framework of *The Commedia*. Initially we find the poet in a tangled and dark wood, symbolic of the sin inherent in nature and man after the Fall. The final terrace of Purgatory presents the dark forest healed and restored to order and fruitfulness—it is the Earthly Paradise, and suggests Eden, the archetypal garden. In Christian lore, prelapsarian Eden was a virtually static paradise, knowing neither seasonal cycle nor climatic excess (a static concept quite foreign to the Chinese mind with its dynamic and centripetal status of the natural cycle). While Dante's Paradise is indeed a place of perpetually temperate climate, perhaps owing to the infusion of Near Eastern fertility symbolism in the early development of Christianity, he also views Paradise within a natural cycle; it is the place from which the seeds of vegetable life in the world below proceed, and to which they return. Yet the progress of the poet from sin to grace does not involve participation in the regenerative flux of nature (as in the Tree of Knowledge blooming in the company of Beatrice and the griffin); it is an allegorical indication of his progress, not so much a "cause" as a "correlate." In many respects the Christian worldview regards divinity as in control of nature rather than revealed through nature. The garden archetype is seen as the conscious image of nature before the Fall; it exists to remind man of his lost innocence and the orderly manifestation of God's will in prelapsarian nature. In essence, as one approaches divinity, the image of Paradise may be seen as allegorically static, rather than naturally cyclic.

Turning from Dante's use of the cyclic aspects of natural symbolism to those of abstract symbolism the reader is immediately struck by the importance of the circle image to *The Commedia*. The turning spheres within spheres, the mandala image of the Rose unfolding, the luminous globe of the sun, the Trinity as three interpenetrating circles of different colours, and of course, the poet's own circular or spiral progress; all are testaments to the power of the circular image in the medieval mind. The most insightful analysis of the metaphysical significance of the circle is undoubtedly Georges Poulet's "The Metamorphosis of the Circle." Concerning himself with Dante, Georges Poulet demonstrates the ontological basis of the circle:

> "Eternity is not simply the pivot around which time turns; it is also that point where, like the rays of the circle, the events of the past and the future converge and unite in the consciousness of God.... Heaven, all nature, the whole of creation in its spatial and temporal unfolding, have existence only because everywhere and always the action from a creative centre causes them to exist. Doubtless this creative action is essentially spatial, since every place in the universe is at the receiving end of the action. But it is also temporal, since every new moment is also the effect of this

continuous creation. God possesses time not solely by His omniscience, but also by His omnipotence."[10]

In simpler terms, God is a sphere of which the centre is everywhere and the circumference is nowhere. The divine concourse of the Holy Trinity is essentially a cyclic interplay which underpins the created universe and moves in ever-widening circles through the universe as continuous creation. For Dante, swift motion, regularly circling like the sun, is a symbol of delight which lies beyond intellectual comprehension. Significantly, the penultimate image of the "Paradiso" is that of the conundrum of man attempting to square the circle; in other words, of equating the earthly (the square) and the heavenly (the circle). The solution is not geometrical but transcendental; the two are indeed one, united "by the love that moves the Sun and the other stars." The circle is then the metaphor of God as eternity, and the paradigm of the transmission of His divine love in continuous creation.

In fact, this brief excursion into the medieval world of *The Commedia* confirms that the cyclic myth is best represented in its mythopoeic aspect as the pattern of the quest of the hero. Without radical difference from the Chinese pattern, the poet-hero progresses from tristia, through itineraria, apotheosis, and finally to reconciliation. It is within the cosmological and existential aspects of the cyclic myth that distinguishing features emerge. Whereas the Chinese cosmos is ideally a cyclic harmony of divinity, man, and nature in accordance with the Tao, the Christian universe is hierarchical. God is the transcendent cause which from **without** preserves his creatures and their individual and continuing existences. Divinity is an "otherness," albeit the support of man and nature through continuous creation, but an otherness which maintains the separate identities of man and nature. Certainly Dante's triune divinity is essentially cyclic but in a sense above and beyond the created world. The Chinese cyclic myth posits the immanence of divinity in nature which is immediately accessible to the seeker of divine reconciliation. Medieval Christianity knew no such immanence in scholastic doctrine, and the historical influences of the more naturally-cyclic Near Eastern cults seem to suggest the reluctance of early converts to accept the estrangement of divinity from nature, and eventually nature from man. It remained for John Milton, several centuries later, to plumb the psychological intricacies of this hierarchical arrangement as it survived in the mind of seventeenth-century Puritanism; but first we must outline the existential revelations of Renaissance humanism which underlay Milton's creation of Satan.

---

[10] Georges Poulet, "The Metamorphosis of the Circle," in *Dante: A Collection of Critical Essays*, ed. John Freccero (Englewood Cliffs: Prentice-Hall Inc., 1965), pp. 154–55.

4. Milton's *Paradise Lost*: A Fall into Linearity

The great victory of the Renaissance was the restoration of divine immanence to the human world. In the words of Georges Poulet: "In a universe which now seemed entirely subject to vicissitude, there remained only a double awareness of the vicissitude itself and the cosmic force which produced it. God seemed rather the indwelling power that from within tirelessly sustained and prolonged the universal motion by which things and beings accomplished their temporal destiny."[11] During the Renaissance, divinity approached the immanent "suchness" of Chinese thought, rather than the "otherness" of the medieval era. In fact, the ephemerality of being in a world of change, the most common initiating theme of the tristia in the Chinese myth, is present everywhere in Elizabethan literature. "Mutability Cantoes" by Edmund Spenser (1552–1599) immediately spring to mind, and William Shakespeare (1564–1616) explored the topic most completely in his sonnets. Shakespeare's response to the fleetingness of being is typically ambivalent: the temporal flux may be viewed negatively as the universal leveller, as in Sonnets XII and LXIV; or positively as the field of action whereby man, especially the artist, could claim immortality through his work, as in Sonnet XIX where the poet concludes: "Yet, do thy worst old time; despite thy wrong / My love shall in my verse ever live young."

Coupled with this ambivalent response to ephemerality, we find renewed interest in cyclical nature as the vehicle of divinity most immediately accessible to man. Already suggested by the allegorical personification in Spenser's *The Faerie Queene*, the use of natural cycles as a model for human conduct finds complete expression in Shakespeare's plays. For example in *The Winter's Tale* the values of nature's regenerative powers are attached to certain characters (primarily Mamilius, Florizel, Perdita, and of course, Autolycus), who move through the play and are instrumental in its resolution. Yet behind Shakespeare's pastoral vision of regenerative nature hovers the famous "degree" speech of Ulysses in *Troilus and Cressida*. Nature was still the lowest link of the Chain of Being, and was corrupted along with humanity by Adam's original sin.

The Reformers emphasized man's fall into a state of corrupted nature and the eschatological concerns of the Redemption and predestination. As a consequence, duration beyond ephemerality consisted of a moment by moment faith in God and personal redemption. Profane nature was more often viewed as a temptation rather than a support, while the efforts of Francis Bacon (1561–1626), Michel de Montaigne (1533–1592), and René Descartes (1596–1650) ended to place man in the position of an observer of natural cycles rather than an equal participant. Seventeenth-century man was a prisoner of the

---

[11] Georges Poulet, *Studies in Human Time,* trans. Elliott Coleman (Baltimore: Johns Hopkins Press,1956), p. 21.

present, his existence preserved moment by moment by an "other" God who guaranteed transcendence of this ephemerality in moral terms at the end of time. The uniquely linear and instantaneous consciousness of man was a favourite subject of seventeenth-century poets, and John Milton (1608-1674) in *Paradise Lost* is no exception. What is most interesting is his encyclopedic rendering of the problem in terms of the entire historical sweep of Christianity.

The three realms of existence in *Paradise Lost*—Heaven, Earth, and Hell, are complex parallels and constantly interacting. We find that the fall of Man parallels the fall of Satan but is also caused by it, and that Man is redeemed by Christ who was the cause of Satan's fall. The pattern of down-fall briefly is this: Satan sins through his envy of the Son's Anointment, then denies his creative union with God in the Son, which in turn results in the false abstraction of spirit into matter and the demonic conceptions of time and space. Satan then causes the fall of Adam and Eve who experience these new modes of time and space themselves.

Even before Satan's physical fall the transformation has begun, as he tries to express through a physical glorification of the self what God expresses through a spiritual glorification of those who form the "one individual Soul." Thus is erected in the North of Heaven the great golden Palace of Lucifer, and later in Hell, the demonic forces build an ostentatious Pandemonium. A second and more important consequence of Satan's rebellion is the new perception of time and space which he gives the fallen angels and in a lesser degree, to man. A hierarchy of perception will make this point clear.

At the apex of the temporal scale is an eternal God; he interchanges past, present, and future tenses at will, and is both in time and beyond time simultaneously. On the next level we find Milton's curious notion of the angelic perception of time. There is a diurnal cycle of sorts in Heaven, but it affects the angels subtly; evening and morning seem to be matters of personal taste. Prelapsarian Adam's temporal experience is, as we would expect, similar to the angels' except his nights are somewhat darker. He has no idea of what death might be and the past is never so important as the present. To the extent that time is illustrated by change through movement, there is no time in Eden.

Fallen Adam's temporal experience is quite different. The separation of mind from God has occurred and Adam becomes acutely self-conscious. The changeless cycles of his previous life have been shattered by discord. The experience of time as a linear progression begins: his past is constantly fading away; the present is never really there, and the future stretches away in "endless miserie." Meanwhile Satan, deservedly, is at the bottom of the scale. He is trapped in a perpetual present of pain and misery. He has no future, absolutely no hope that there will be any change in his condition. Of course there is hope for Adam (though not realized until the concluding lines of the epic) in the cycle of life and death in Christ's redemption.

There is a similar scale of identification in the experience of space. For God, space is eternal presence. He is everywhere at once, and the identification is total. For unfallen Adam and the angels, space has definite limits and within this circumscribed area all is right. They cannot "be" everything but they can know that everything is in its proper place. Adam especially delights in participating in the natural harmony of Eden until the Fall, when space becomes an indifferent environment. There are blustering winds, varying seasons, and the animals have become carnivorous. Adam is terribly disordered and alone but he is not threatened. Satan has naturally fared worst. He experiences the completely alien and hostile environment of confining Hell without and the chaos of insanity within.

The last two books of *Paradise Lost* enable Adam to transcend the present and see mankind's hope for reunion with God at the end of time. Satan must live in the present because he cannot die; only eternal damnation awaits him. Paradoxically, man's corruption is not eternal because God has granted him death, and it is in Book Eleven that Adam is first familiarized with death. As the visions unfold, he is acquainted with all other facets of behaviour, both good and evil, which comprise the fate of his offspring. Adam becomes humanized by suffering as a parent all the consequences of his disobedient act. But as he remains unconvinced that his fall was truly "fortunate," the true plan of God's infinite mercy is finally revealed when he discovers that Christ will give his own life so that mankind may be redeemed from death into an eventual reunion with the creative force of God in Heaven. At last Adam understands the meaning of history, a history in which his progeny will enjoy the reunion that he has forfeited. Michael's last promise is of the final victory over Sin and Death, the Last Judgement, at which time all hierarchies will dissolve into the Pure Light of God. Finally our general parents pass into historical time, secure in their faith, and strengthened by an understanding of the interactions between time and eternity.

In *Paradise Lost* then, Milton presents the threefold legacy of Adam's fall in terms which circumscribe seventeenth-century man's consciousness: in time, existence which endures from moment to moment by virtue of continuous creation; in space, existence which is alienated from a nature which is itself corrupted; and finally, a linear concept of a history which provides possible divine reconciliation only at the end of time itself. To be sure the pastoral delights of cyclical nature were never truly inaccessible, but there is little sense of revealed divinity or the possibility of permanent harmony with nature as a path to divinity. Rather, survival of pastoral conventions represented a positive response to the present, a moment of joy. Typically the negative response is almost as frequent and usually concerns such archetypal images as the reaper, mower, or scythe. In essence, the eschatological emphasis of Reformed Christianity rendered the regenerative powers of the cyclic myth to an individual moment of consciousness; lasting harmony was relegated to the Last Judgement well beyond the natural sphere.

5. Goethe's *Faust*: A Striving for Meaningful Duration

During the eighteenth century the retreat of divinity from the role of continuous creation accelerated. According to the principles of Cartesian detachment, the universe came to be perceived as a complicated mechanism of secondary causes originally set in motion by God, the primary cause. In this rationalist view man should look to nature not for the noumena of pantheistic experience, but for evidence of the intelligent design of creation which it presents. As the subject-object viewpoint tended to estrange man from nature emotionally, and as divinity was no longer the support of existence, self-consciousness came to be instantaneously preserved by sensation alone, while memory served to string the individual beads of experience into a sense of individual duration. In the most optimistic view, lived sensation (however momentary) became the consciousness of being; surely the apex of this "interiorization" was reached with Rousseau's proclamation, "Like God one is sufficient unto oneself."[12] Indeed, if Jean Jacques Rousseau (1712-1778) in *Confessions* of 1728 represents the ultimate "interiorization" of consciousness in the early eighteenth century, the work must also be acknowledged as the foundation of Romanticism which developed fifty years later in England and Germany. Eighteenth-century man was conscious only in a moment of sensation; yet as the moment became ever more imbued with feeling, the exceptional man might perceive a self and a reality which were not instantaneous, in short a duration based on a profound revelation. This revelatory moment (whose consequent loss is the ever-present Romantic nostalgia) might be based on communing with nature, or the transcendent loss of self in love, or a feeling of the entire past reborn within oneself. Whatever its cause, this Faustian striving for a duration beyond the moment became the essence of pre-Romanticism and certainly a central concern of the drama of Johann Wolfgang von Goethe (1749-1832).

The presentation of heaven as a cosmogonic reality is a central assumption of *The Commedia* and *Paradise Lost*, one that is significantly absent in Goethe's *Faust*. As a scientific investigator and poet-philosopher, Goethe's fundamental purpose was to represent, in a work of art, man's place within the confines of life on this earth; to explore man's potential and limitations within the natural laws to which he is subjected. In *Faust*, heaven exists for dramatic purposes; earth (and the poet's mind on earth) is the real field of action. Goethe's scientific and philosophical researches were directed towards the "universal"

---

[12] Cf. Poulet, *Studies in Human Time*, p. 21.

which underlies phenomena.[13] To this end he embraced the Neo-Platonic idea of a world-soul from which life emanates and to which it ultimately returns, an image of the overall harmony into which unharmonious life on earth might finally resolve. If life is considered to be a series of emanations issuing from and striving to return to the living godhead, it follows that all matter is in a state of continuous flux and change. Moreover as the unceasing cycles of the world-soul animate all physical life in perpetual change, value concepts of durability and permanence find no true correspondence either in nature or the life of man. As the play vibrates with a vitality which can find no satisfaction in any static situation, so Faust strives for a duration beyond the eighteenth-century moment of "sensitive" consciousness.

Faust's cynicism regarding the possibility of permanent value within natural flux underlies his bet with Mephistopheles. He tells the devil that if he should ever wish a beautiful moment to last, then he will be ready to die and surrender his soul. Paradoxically, by signing the pact Faust opens the floodgates to previously denied emotional and sensual experience, the romantic basis of the revelatory moment and the intuitive understanding sought by his "higher" soul.

In *Faust* time and space become functions of the imagination as the seeker discovers that every experience is transformed into illusion as reality recedes within the universal metamorphosis. These contrasts between illusion and reality and between form and change comprise the rhythm of cyclic alternation according to which the play functions. From the opening scene where the angels present a view of earth in endless rotation dominated by the quotidian cycle, to the continually shifting landscape of the final "Forest, Rocks, Solitude" scene, Goethe demonstrates the universality of cyclic alternation in human experience. In Faust himself the rhythm of day and night symbolizes the struggle between biological desire and spiritual aspiration. Thus in the "Night" scenes which are the first and last appearances of Faust, the first is preceded by the angelic scene mentioned above which is bathed in light, and the last is succeeded by a lighted scene in which Faust's immortal spirit is carried upward.

A second focus of cyclical natural symbolism is Goethe's frequent use of water imagery in all its forms. With its ever-changing quality, its perpetual motion, and the fact that it is a life-giving force, the water cycle affords an ideal image of the Neo-Platonic world-soul, which similarly emanates into matter and then returns to itself. Goethe makes this point most clearly when Thales says to the Ocean: "From the wave was all created / Water will all life sustain / ...Your

---

[13] As Harry Slochower points out, "Long before Fichte and Whitehead, Goethe opposed what Hopkins calls 'the bifurcation of nature'. His 'symphonistic method' aimed to find the recurrent phase, the primal phenomenon or 'Urphanomen'; in botany, the 'Urpflanze', in the animal world, the 'Urtier'. The key words in his vocabulary are 'world' and 'unity'." Harry Slochower, *Mythopoesis: Mythic Patterns in the Literary Classics* (Detroit: Wayne State University Press, 1970), pp. 187-88.

power the freshness of our life maintains."[14] While water is the poet's favourite image of metamorphosis, we also find references to the evolution of a butterfly from its chrysalis, the unfolding of the leaf from its bud, and of course, more mythical images of death and rebirth.

Goethe's use of natural symbolism is strongly indicated in the events surrounding his apprehension of the Earth Spirit very early in the drama. Unable to partake of the serene yet overwhelming vision of the Macrocosm, Faust's magical powers conjure the Earth Spirit which combines water and Neo-Platonic imagery; it is a concrete representation of the dynamic force underlying all change (in a sense, similar to the Chinese Tao.) Unable to grasp the Earth Spirit's significance, a dejected Faust is on the verge of suicide when the Earth Spirit remanifests itself in the Christian rebirth symbolism of Easter bells which Faust connects with youthful memories of spring festivals. This celebration of the cyclical regeneration of nature and man moves Faust to tears, and compels his descent to the valley of experience. Such cyclical symbolism represents the flux of eternal return in the poet's world, but only rarely provides a sense of being beyond momentary sensation. The vitality of Goethe's nature admits no permanence beyond the mortally inaccessible world-soul.

Whereas Dante's work summarized some thirteen centuries of Christianity, Goethe's embraces nearly all of Western culture and the myths of three thousand years. The cyclical aspects of *Faust*'s myth symbolism would alone fill several volumes and we must restrict ourselves to its most salient features. Hovering in the background is of course the great myth of Christianity, and one of the strengths of *Faust* is its use of Christian doctrine as myth, still a heretical approach in the eighteenth century. Ignoring Christian propaganda, its imagery provides a mythical setting in which the drama unfolds. Classical mythology forms an integral aspect of the work, especially in the Helen sub-plot and the Classical Walpurgisnacht. In particular the latter festival concentrates many of the Dionysian and fertility figures of antiquity to emphasize the erotic and sensual side of Faust's "lower soul." The regenerative aspect of the natural cycle provides for the rebirth of Homonculus at the oceanic beginning of time and the path to the corporeal Helen, as opposed to her conjured shade. Here the myth of the birth of Aphrodite from the sea is presented, and transformed into the cyclic myth of Galatea. As the representative of Eros, Galatea's reunion with Nereus prefigures the reappearance of Helen in the next act and Gretchen after Faust's death, and certainly the culmination of the Christian myth at the play's conclusion. In short, Goethe freely draws upon the most potent myths of eternal return to underscore the perpetually regenerative aspects of the flux of reality, and to suggest the final redemption of Faust himself.

---

14 Goethe, *Faust: Part Two*, trans. Philip Wayne (Middlesex: Penguin Books Ltd., 1959), p. 154.

In one of the most striking examples of Goethe's 'modernity', Faust penetrates into the shadowy world beneath myth, the realm of the Mothers. In essence the Mothers guard the memory of mankind, and represent an elemental maternal instinct at the deepest levels of the human psyche. One is tempted to suggest that they are Goethe's pre-Jungian version of racial memory, but the vagueness of description ("eternal pastime of the eternal Spirit") only indicates that their transforming powers are related in some way to the natural flux. Perhaps Faust's journey to the Mothers is best viewed as an attempt by the mythic hero to apprehend the secret of nature by retracing his way to the origins of creation. Inasmuch as the journey brings him first to Helen of Troy and finally to the Mother of God, it is successful.

In Part One Goethe demonstrates how Faust's titanic mission destroyed the innocent lives of Gretchen and her family, while in Part Two Faust's great social projects entail tragic social sacrifices. After the hero has been blinded by Sorge (or Care), he feels an "inner light" which compels his final vision of a social Utopia and his blissful acceptance of the fleeting Moment. Despite Mephistopheles' excitement, Faust is not declaring his desire for the static moment to endure (to recall his original bet); rather, as Harry Slochower indicates in his *Mythopoesis*, "the words salute a continuous task in which the 'Moment' is forever renewed."[15] The key is not in mere activity which seizes the moment, but in the direction of the activity (and here Goethe is most anti-romantic), which is directed towards the betterment of mankind. In *Faust* we see the resurgence of the cyclic myth, generally within a Neo-Platonic and classical framework, which allows for progress or evolution within the natural cycle as a dialectical upward striving. The revelatory moment is seized only that it may be renewed by further striving. Reconciliation with divinity is possible (we do not know whether Faust actually reaches Heaven), but the rewards are most clearly evident in earthly existence. Within the cyclic manifestations of the world-soul, duration beyond the moment is acknowledged, in a sense "synthesized," then renewed in the next moment at a more evolved level. That such a progression is available only to a "super-hero" upon his death is a final testament to Goethe's personal doubts regarding the perfectibility of man.

6. Shelley's "Ode to the West Wind": A Poet's Vision at the Cycle's Centre

One would expect that the Romantic movement would provide the definitive reintegration of the cyclic myth with nineteenth-century literary concerns. Certainly on the simplest level the Romantic poets rehabilitated and rediscovered nature on a scale unmatched in the last millennium; yet so acute was the self-

---

[15] Slochower, *Mythopoesis: Mythic Patterns in the Literary Classics*, p. 209.

consciousness of the moment, so intense was the Cartesian legacy of "interiorization" of reality and the Faustian striving for possession of reality, that the urgency of the goal almost dictated its evanescent apprehension and consequent loss. As continued scientific advances emphasized the mechanical design of celestial nature, terrestrial nature and man himself became the surviving field of divinity. As a vital being within an organic world, man's relationship to nature came to be seen as one of participation, an identity of process rather than a separation of subjective and objective creatures or products. A revived sense of the numinous power of nature became symbolically related to human nature; whatever is immanent in nature is also imminent within ourselves, and this is what distinguishes Romanticism from mere pantheism, a point well made by Samuel Taylor Coleridge (1772-1834) in his "Ode To Dejection:"

"O Lady! We receive but what we give,
And in our life alone does Nature live."[16]

But if nature in some way represents man, one must ask what is the nature of this representation? Here the English School of Romanticism recognized the "imagination" as the central identity between man and nature. The influence of German philosophy was crucial; Immanuel Kant (1724-1804) had proved that reality is forever unknowable to the intellect, and Johann Fichte (1762-1814) that nature itself was only a creation of the ego or the private "will." Denied an intellectual identity of man and nature, the English Romantics seized upon the imagination which participates with nature as a process, and imitates specifically its power of bringing organisms to birth. Coleridge, in particular, determined the poetic imagination as the "exemplastic" power which reshapes our primary awareness of the world into symbolic avenues to the theological. That vague yet vital force which drives nature through cyclic regeneration even in its most violent manifestations is also to be found within the human psyche; the contraries of nature are also those of man.

A brief examination of a representative poem, "Ode To The West Wind" by Percy Bysshe Shelley (1792-1822), will illuminate the relationship between cyclic nature and man as conceived in one of its more profound studies. The poet at sunset observes the turning of the year, the passage into fall. As night comes on, a violent tempest of hail and rain descends, an event which the poet reads as a sign of the creative destruction that will affect the whole condition of man. In the first stanza, Shelley establishes the "west wind" as the breath of life, the source of seasonal change; yet it cannot be apprehended directly. The poet may only observe its manifestations, and distance himself from the source

---

[16] Cf. Arthur M. Eastman, ed., *The Norton Anthology of Poetry* (New York: W. W. Norton & Company Inc., 1970), p. 611.

by describing its effects. The vitality is both destructive and creative; thus in a state resembling death, the seeds as symbols of renewal lie in a bed (sexual energy), until the wind infuses them with life as the "breath" of the first line becomes a stronger "blow." The stanza concludes with the first of several invocations to the wind to hear its prophet, the poet:

> "Wild Spirit, which art moving everywhere;
> Destroyer and preserver; hear, oh, hear!"[17]

In the second stanza Shelley transfers his gaze to the sky. Here the heavens, like the forest, are dying, and yielding their substance up to the destructive wind. The compression of time from a year to a night signals the nadir of the destruction-preservation cycle in theological imagery, to reinforce the idea that this now destroying vitality is the actual realm of divinity.

> "Of the dying year, to which this closing night
> Will be the dome of a vast sepulchre,
> Vaulted with all thy congregated might
> Of vapours, from whose solid atmosphere
> Black rain, and fire, and hail will burst: oh, hear!"[18]

With the third stanza Shelley's social concerns are most evident as the wind destroys what is left of the old "Mediterranean" order of "old palaces and towers / quivering within the waves intenser day." By definition, the wind is also the creator of this order, but it is the mind "lulled by the coil of his crystalline streams" that supports itself by restructuring the manifestations of the west wind, which lends legitimacy to the dying order despite the fact that the reflected civilization is overgrown with vegetation.

In the last two stanzas the poet himself replaces leaf, cloud, and wave as the object of the wind's force. In the fourth he declares that if he were merely part of nature (like a leaf or cloud), or if he still possessed the imaginative strength of childhood, he would not be striving in prayer for communion. That both conceptions are impossible is witnessed by the "fall" of the poet upon "the thorns of life." Denied apprehension of the life-force by self-consciousness and the linearity of time, Shelley, in the fifth stanza, turns to the earlier suggested identity between the poetic imagination and the vitality behind nature. The poet demands to be like a "lyre," the perfect image of art compelled by nature; he wishes to be an instrument like the forest, sensitive to cyclical changes yet persisting through those changes. A cycle is implied as Shelley's thoughts

---

[17] Cf. Eastman, *The Norton Anthology of Poetry*, p. 657.
[18] *Ibid.*

("dead" because of their inherent "selfishness") may become structures which support new life, in the same manner as a flower bed may spring to life when infused with breath. The prayer to the wind stresses mutual need; if the prophet needs the divine, the divine as assuredly needs the prophet if the message is to be heard by men. The "lyre" has become a "trumpet"; the wind, as winter becomes spring, must blow through the poet rather than around him.

For all its revolutionary intents, Shelley's prophetic vision was essentially a private vision, as were all Romantic visions. In his introduction to *The Visionary Company*, Harold Bloom notes that because Lord Byron (1788-1824) was the most social of Romantic imaginations, he was the least Romantic.[19] The use of the cyclic myth as a structural model for art, and cyclic imagery as the immanent force behind both nature and the poetic imagination, was never so widespread as during the period of Romanticism. Inasmuch as the poetic imagination is a private imagination, resurrected and maintained by a creative elite, apprehension of the unity between man and nature was reserved at first for the most imaginative. But one can imagine Wordsworth's pastoral shepherds trampled in the rush of those seeking communion with nature in the years to come. Socialization along Cartesian and materialist lines rendered "original" participation with nature inconceivable; "imaginative" participation became popularized to the point of meaninglessness.

In China the cyclic myth has survived simply because it began as a myth (or complex of myth) which evolved an accepted cosmology reflected in ritual and art, in government and individual life. The failure of the French Revolution may be seen as the failure to create a new myth, that of the recently discovered "natural man." The Romantic movement was an intellectual revolution which cannot be faulted in its attempt to provide the artistic structures of that new myth doomed to failure because of its "newness."

The legacy of the Romantic movement was the progressive "interiorization" of nature, witnessed in the rise of the science of psychology and the riot of private symbolism in both visual and verbal art, and the subsequent absence of a divinely-ruled cosmogonic order as a background universal in art. Since the Romantics, the popular experience of time has remained one dominated by linearity and the linear progress of history. By way of contrast, we recall that the archaic mentality defended itself against linear history by periodically abolishing it through repetition of the cosmogony and a periodic regeneration of time, or by giving historical events a meta-historical significance that was consoling and coherent with the cosmic order. How different from this cyclic mentality is the post-Hegelian position, in which every effort is directed toward saving and conferring value on the historical event as such, the event for itself

---

[19] Cf. Harold Bloom, *The Visionary Company* (London: Faber & Faber Ltd., 1961), p. xv.

and in itself. As existence becomes more and more precarious because of history, the position of linear historicism becomes less and less consoling. With this problem in mind, in the past century we have witnessed attempts by many artists to rehabilitate the cyclic view of history, most notably James Joyce and William Butler Yeats.

7. Yeats' *A Vision*: A Prophecy about Cyclic Cosmos

Despite the dominance of Christian eschatology in Western thought, the cyclical ideology was never totally overwhelmed in intellectual circles, and certainly not in cultures which maintained an agrarian base. Beside the conception of linear progress, it survived in the astronomical theories of Tycho Brahe and Johannes Kepler, and more significantly in the historical theories of Giordano Bruno, Giambattista Vico (1668–1744), and most recently, Oswald Spengler (1880–1936). The cyclic conception of Vico is especially relevant in modern literature as it influenced both William Butler Yeats (1865–1939) and James Joyce (1882–1941) in their temporal thought.

The herculean efforts of Yeats to rehabilitate the chaos of modern history into a coherent cosmos represent one of the great intellectual adventures of this century. Over a period of nineteen years Yeats explored the occult imagery of the Kabbalah, Neo-Platonism, and the hermeneutic texts of the London theosophists. Aided by his wife's phenomenal psychic powers of "automatic writing" he finally completed his work, if completion could be possible in such occult areas, and in 1937 published *A Vision*, his personal vision of the true arrangement of reality. Developing his earlier conception of overlapping gyres, Yeats believed that the image of reality was best a Great Wheel of twenty-eight spokes, a lunar cycle corresponding to the phases of the moon. The Great Wheel, representing everything, rendered a coherent reality beneath the chaotic flux of mundane experience, and was categorized through the twenty-eight personality types, the twenty-eight incarnations a man must live through, the twenty-eight phases of any single life, and the twenty-eight basic phases of each two-thousand-year cycle of world history. Such an esoteric scheme brings us very close to the original concerns of the cyclic myth, especially in its explicit identification of man with the cosmos. The following passage, from Yeats' *Wheels and Butterflies*, makes this point more directly than we could suggest through commentary.

> "...the soul realizing its separate being in the full moon, then, as the moon seems to approach the sun and dwindle away, all but realizing its absorption in God, only to whirl away once more: the mind of a man, separating itself from the common matrix, through childish imaginations, through struggle—Vico's heroic age—to roundness, completeness, and then externalising, intel-

lectualising, systematising, until at last it lies dead, a spider smothered in its own web: the choice offered by the sages, either with the soul from the myth to union with the source of all, the breaking of the circle, or from the myth to reflection and the circle renewed for better or worse. For better or worse according to one's life, but never progress as we understand it, never the straight line, always a necessity to break away and destroy, or to sink in and forget."[20]

Immediately evident is Yeats' ultimate denial of linear progress, to be replaced by a cyclic conception of the identity of the soul, the true man, with cosmic regeneration.

The cyclic cosmogony of *A Vision* presented Yeats with a system of metaphor which frequently is at the heart of his most notable poetry, "The Second Coming" being the most famous example of this method. It is a credit to the poet's genius that his unique "vision" is neither didactic nor obscuring in his art; as Richard Ellmann comments, "an awareness of the system was more useful for writing than it is for reading the poem(s)."[21] Yet for all the poet's labours in the labyrinthine lore of the occult, for all his genius and personal popularity, his private "cyclic myth" remained a "myth" in its most profane sense of "fiction"; in the mass mind of the twentieth century, time is linear, history progressive, and the poet's prophecy singularly unheard.

8. Joyce's *Finnegans Wake*: A Viconian Myth of the Course of Humanity

Allied with Yeats against the progressive linear conception of history, in *Finnegans Wake* James Joyce declares history to be "A human pest cycling (pist!) and recycling (past!)..."[22] For Joyce, not only does history repeat itself, but "the Vico road goes round and round to meet where terms begin."[23] The Viconian myth of history is to *Finnegans Wake* what the Homeric myth was to *Ulysses*: a structural device of complex parallels which tends to invest the microcosm of character with vast and profound echoes of the macrocosm. Harry Levin's critical introduction to the works of Joyce offers a concise summary of Vico's theory which we present in the interests of brevity.

---

[20] Cf. William Butler Yeats; cited by John Unterecker, *A Reader's Guide to William Butler Yeats* (New York: Noonday Press, 1959), p. 254.

[21] Richard Ellmann, *Yeats: The Man and the Masks* (New York: E. P. Datton, 1948), p. 233.

[22] James Joyce, *Finnegans Wake* (New York: The Viking Press, 1939), p. 99.

[23] Joyce, *Finnegans Wake*, p. 452.

The Viconian pattern of repetition presents "three consecutive periods... characterized as divine, heroic, and civil. Each period contributes its characteristic institution (religion, marriage, and burial rites) and its corresponding virtue (piety, honour, and duty). The inarticulate dark ages give way to the fabulous, and then the historical, forms of literary expression; the original hieroglyphic language is succeeded by metaphorical speech, and at length by an epistolary style and a profane vernacular. The rise of cities is the sum of three epochs of man's activity, yet the ruins of bygone civilization foreshadow the fall of cities. The fourth epoch, and the peculiar twist in Vico's philosophy of history, is the cyclic movement by which the third period swings back into the first again 'da capo'."[24]

In *Finnegans Wake* the recurrent domestic cycles of *Ulysses* are subsumed within the larger cyclic matrix of Viconian philosophy. Thus the first three books of the novel represent divine, heroic, and human ages respectively, while the fourth is the reflux that leads to the re-creation of the divine age again. Moreover, in Book I the cycle is twice repeated on a smaller scale in the first eight chapters, and in Books II and III it is again twice repeated in the four chapters of each. The main structure is visible as one large cycle, containing four smaller cycles. This reductive tendency is evident as "wheels within wheels"; cyclic images and incidents abound on almost every page of the work. As Levin notes, "The whole sequence is likely to emerge coloured by its context at any moment: '...their weatherings and their marryings and their buryings and their natural selections...' or as '...thunder-burst, ravishment, dissolution, and providentiality...' and again, '...eggburst, eggblend, eggburial, and hatch-as-hatch can...',"[25] The effect of such a technique is to suggest a world of cyclical immanence, where myth and arcane correspondence hover behind every gesture or turn of phrase.

Many of the cyclical sub-plots and incidental events are conceptually independent of Vico's theory, yet in the final analysis generally correspond to the cyclic structure. The Earwicker family is the family of man moving through birth, death, and rebirth. The first four chapters follow the father through these phases, and they are represented again in the next four in the cycle of the mother, from her original letter to her renewal as the river. Book I concerns the father, the mother, and Shem, her favourite son and his successor. Finally in Book IV, the mother (again as a river) renews herself and family once more.

---

[24] Harry Levin, *James Joyce: A Critical Introduction* (New York: New Directions Publishing Co., 1941), p. 145.

[25] Levin, *James Joyce: A Critical Introduction*, p. 148.

Within the structural framework afforded by Vico's regenerative spiral of history, Joyce creates a pan-cultural mythological complex which borrows freely from Scandinavian, Oriental, Germanic, Celtic, Romanic, and of course, classical and Christian mythologies. According to John Vickery, the encyclopedic inclusiveness of *Finnegans Wake* suggests more than *The Golden Bough* by Sir James George Frazer (1854–1941). Thus the novel is in effect "a human comedy on man's religious consciousness, dramatizing a secularized and comic version of the struggle between the religious guilt and fear and the imaginative satisfaction and sexual joy. Informing the fear is the fact of mortality. By stressing the ritual forms of death, *Finnegans Wake* follows closely the lead of *The Golden Bough* and is particularly sedulous in using its images and figures."[26] John Vickery goes on to outline the extensiveness of Joyce's Frazerian images, focusing on the interplay between Shem and Shaun. It must be remembered that whereas Frazer's work collects, arranges, and comments upon the myths of archaic man, he does not presume their continuing operation within a modern consciousness. The sheer inclusiveness of Joyce's culminating work (W. Y. Tindall calls it "perhaps the most comprehensive of myths"[27]) poses problems which are implicit in the mentality of the twentieth century. Indeed Yeats and Joyce's attempts to rehabilitate the cyclic view of cosmos and history are the last and greatest intellectual responses to an alienation of cyclic mentality for more than three centuries from Defoe, Richardson, and Sterne to Leonard Cohen and Samuel Beckett.

---

[26] John B. Vickery, *The Literary Impact of the Golden Bough* (Princeton: Princeton University Press, 1973), pp. 410–11.

[27] Cf. W. Y. Tindall, *James Joyce: His Way of Interpreting the Modern World* (New York: Charles Scribner's Sons, 1950), p. 103.

B. ASPECTS OF TIME IN EARLY ENGLISH NOVELS: *MOLL FLANDERS, PAMELA,* AND *TRISTRAM SHANDY*

1. The Philosophical Background

Time is the essential experience of our existence. In contemplation, we do not find ourselves extending through space, but rather, enduring through time. It is this implicit correlation between consciousness and time which has plagued analytic investigations of existence. The traditional ideas of time, involving concepts of past, present, future, and eternity do not adequately describe either ideal or experienced time. They are at best convenient terms of reference whose utility disappears upon close examination. With the concept of external time, the relation between past, present, and future is a confusion of time with space evident in the usual metaphor of the time-line and in such terms as "hereafter" and "thereafter." Within consciousness, the dream state exhibits a confusing mixture of past, present, and future in which no boundaries can be discerned. In the mystical state, one supposedly communes with eternity, yet this word is defined as "timeless." In light of these conflicts and ambiguities, we shall refer to these usual terms as only one aspect of time. Insofar as the evolution of the novel has demonstrated an increasing awareness of the content and processes of consciousness, we shall begin with the formulation of a temporal framework which will circumscribe both ideal and experienced time.

In general time is never experienced as a whole; we find that what was once the future has become the present and will soon be the past. In this view time becomes a linear succession, but succession alone can not account for the fullness of time. As a progression of present moments, pure succession denies the reality of past and future which must be inherent in the present for anything to exist. The fact that things exist through time demands a conception of permanence: duration. Because duration is eliminated from the purely successive view, it can only be symbolized by extra-temporal concepts. Thus duration is relegated to a space-time continuum, or even to the state of temporal transcendence, eternity. If we approach time only from the standpoint of duration, where time is but an actual present, then the reality of past and future is questionable. The failure of each of these theories to explain time lies in their common notion of past, present, and future as forms of the present; in other words, time is to be understood as either a succession of present moments or else as an all-inclusive present. The dilemma of this either-or proposition may be resolved within a synthetic formulation which yields a more accurate view of time.

We have seen that without succession, duration must be an unchanging present; thus as an adequate description of time, duration may arise only from the flux of time. Conversely, only within the background of duration are the emergence and our awareness of succession possible. Duration then becomes not a static but the sustaining quality of time, and therefore the actual "co-existence" of past, present and future, a "co-existence" which does not imply simultaneity because succession is included in their relationship.

The concept of duration has important ramifications for consciousness; the individual as a self-centred form of life must endure by opposing its permanency to its changeability in time. The highest state of consciousness, the transcendental, is not "timeless" within our formulation. Rather it appears as an awareness of the support of duration, the awareness of the "co-existence" of past, present and future, or as the experience of time as a whole. In the dream state, the usual order of succession in terms of cause and effect is suspended and replaced by associations of an unconscious or even mythic nature. In temporal terms the waking state of consciousness may be described as primarily the experience of succession with duration implicit in the continuous perception of self-identity. Novelists have long been aware of the difference between the experience of succession in consciousness and the pure succession of clock-time or the time-line. Memory exhibits a non-uniform dynamic order of events. Wishes and fantasies may be recalled as facts; and facts remembered are modified and reinterpreted in the light of present circumstance, past fears, and hopes. While memory demonstrates the usual temporal qualities of succession and duration, its structure seems to be determined by significant association rather than causal connections. With the above examples in mind, the utility of the succession-duration approach should be evident. Every level of consciousness posits a slightly different "mix" of succession and duration. These two terms will define our investigation of the shifts in temporal thought from the Renaissance through the seventeenth and eighteenth centuries, and ultimately their importance in the rise of the English novel.

During the Renaissance, it was believed that the linear progression of "becoming," as exhibited in nature, was transformed and sustained by divine infusion. Within the chain of existence, "each being appeared to be an autonomous entity which found in itself the inexhaustible resources to engender its continuance by a diversity of motion."[1] This divine force was immanent in the object itself; a spontaneous intercommunication in all individual activity within the cosmic "becoming" was permanently felt. Thus duration and existence were nearly identical, and human thought saw itself as part of all things.

The seventeenth century was the epoch in which the individual discovered that his duration was different from the Renaissance ideas of both succession and duration. With the new rationalism, human thought distinguished itself from the successive order of nature in order to reflect upon it. This distancing process was evident in Francis Bacon (1561–1626): "Those, therefore, who determine

---

[1] Georges Poulet, *Studies in Human Time*, p. 9.

not to conjecture and guess, but to find out and know... must consult only the things themselves."[2] Man could not be upheld by this force in nature. No longer a part of things, he was always aware of his separate existence.

Cartesian dualism was also responsible for a shift from Renaissance attitudes towards duration. At times the Renaissance man felt himself to be the creator of his own destiny. Duration could be a field of action in which, despite his mortality, man could reveal his authentic divinity, and gain a personal immortality. This spirit was evident in many of Shakespeare's sonnets.

> And yet to times in hope my verse shall stand
> Praising thy worth, despite his cruel hand.
> (Sonnet LX, 13-14)

In contrast, the seventeenth-century man, exhibiting Cartesian detachment, discerned that duration was only a successive progression of thoughts, a flickering lifetime of instantaneous **cogitos**. He experienced an acute sense of the discontinuity of duration. The result of this perception was a dependence upon God's continuous creation. For if existence was limited to a fleeting moment, then man's life should be somehow prolonged from instant to instant. During the Renaissance, continuous creation emanated from God and supported man in an existence which was identical with duration. In the seventeenth century, existence and duration became separated and overlapped only in an instant. Thus man's existence was sustained from instant to instant and if God should withdraw his support, even for a moment, man would cease to exist.

During the eighteenth century, divine support of duration became even farther removed from everyday life. God as the first and continuing cause of existence had been replaced by feelings, sensations, and whatever caused sensations. The Cartesian cogito gave way to that of John Locke (1632-1704), "I feel, therefore I am." The abysmal nothingness which separated instants was the nothingness of pure insensibility, and to escape this negation was to be aware of one's own sensations. The more intense they were the more one would feel his present existence; and the more numerous they were the more one sensed a duration in his existence. In a world which had come to be regarded as a complex mechanism set in motion aeons before, it seemed natural that the invention of the clock should occur, and that its use should spread. As Hans Meyerhoff notes: "The philosophical notions of clear and distinct ideas (René Descartes) or of simple and separate impressions (John Locke and David Hume) may now be looked upon as being constructed after the model of temporal units in physics... or rather, the analysis of experience into sensations and impressions

---

[2] Cf. Bacon, cited by Basil Willey, *The Seventeenth Century Background* (New York: Doubleday & Co., 1953), p. 33.

is analogous to the analysis of the temporal continuum into distinct, measurable units."[3] The tremendous shift in temporal attitudes, in which multiplicity of sensation replaced God as the support of duration, was best summarized by Jean Jacques Rousseau (1712-1778): "Like God one is sufficient unto oneself."[4]

The eighteenth century was also the period in which man emphasized "memory." Each new moment of sensation revealed not only the kernel of the moment, but also the mass of sensations already lived. By remembering, man escaped the purely momentary, the nothingness which lies between moments of existence. To exist, then, was to be one's present, and also to be one's past and one's recollections. The rise of memory was important in a more general sense, for it was paralleled by increasing awareness of the depth, order, and direction of human history evident in the Neo-Classicism of the period.

One cannot help but discern the influence of Locke in these new temporal attitudes. His thought lies behind the idea of sensation as the actions of reality upon the mind, and memory as the "combination" of these sensations. Locke believed that certain ideas come to be associated in a man's mind, either by their natural correspondence, by chance, or by custom, and that such associated ideas "always keep in company, and the one no sooner at any time comes into the understanding, but its associate appears with it; and if they are more than two united which are thus united, the whole gang, always inseparable show themselves together."[5] The momentary appearance of this "gang" assures the individual of his continuity, ensuring a duration without divine support.

The influence of French temporal thought on the English mind is nearly as important as that of Locke. The English philosopher proposed the notion of sensation and the French magnified it to excess. As Georges Poulet notes: "The man of the eighteenth century wants to accentuate at any price the intensity of the moment of feeling: either by a sudden reversal of situation... or the sadistic instant."[6] Thus Rousseau proclaims: "Ah! How happy we would be if heaven would annihilate the tedious intervals which retard the blissful moments."[7] While his friend Étienne Bonnot de Condillac (1715-1780) stated: "To live is properly to enjoy, and that man lives longest who is most proficient at multiplying the objects of his enjoyment."[8] The sensual splendour of the French court, the profuse sentimentality of the French nobleman, and more generally,

---

[3] Hans Meyerhoff, *Time in Literature* (Berkeley & Los Angeles: University of California Press, 1968), p. 90.

[4] Cf. Rousseau, cited by Poulet, *Studies in Human Time*, p. 21.

[5] John Locke, "An Essay Concerning Human Understanding" in *Great Books of the Western World*, Vol. 35 (Chicago: Encyclopedia Britannica, Inc., 1952), p. 249.

[6] Poulet, *Studies in Human Time*, p. 22.

[7] Cf. Rousseau, cited by Poulet, *Studies in Human Time*, p. 23.

[8] Cf. Condillac, cited by Poulet, *Studies in Human Time*, p. 24.

the broad sense of "joie de vivre" made their way into English life and are satirized in many early novels as the vaguely French fop. Lord Bellarmine in *Joseph Andrews* by Henry Fielding (1707-1754) comes readily to mind.

In summary, the eighteenth-century man experienced life temporally as a successive series of moments strung together and supported by a duration based on personal memory and multiplicity of sensation. Each moment was not only one of sensation but a chance for feeling, and the depth of feeling was, in a great part, responsible for the value of the moment, the value of one's life. Before dealing with several representatives of the early English novel, it is necessary to suggest a framework within which a novel may be approached from the standpoint of time.

First, whereas life displays both succession and duration, language is limited to expression of the purely successive. It cannot express simultaneous experience nor the flow of consciousness. It remained for the "stream-of-consciousness" novelists at the turn of this century to tackle this problem, and thus our discussion will not focus on the depiction of mental life in the early novel, but on the form of these novels, and how their structures and themes are coloured by eighteenth-century temporal attitudes. Ian Watt's concept of "realistic particularity" is especially helpful in this regard. He believes that the primary implicit convention of the early novel was "that the novel is a full and authentic report of human experience, and is therefore under obligation to satisfy its reader with such details of the story as the individuality of the actors concerned, the particulars of the times and places of their actions, details which are presented through a more largely referential use of language than is common in other literary forms."[9] The importance of "detail" is also the importance of the successive moment, its content of sensation and feeling; in other words, the novel's formal realism was in part dictated by the prevailing attitudes towards time.

2. Defoe's *Moll Flanders*: Memoirs of Sensory Details of Successive Moments

Speaking of Daniel Defoe (1659-1731) and his achievement in *Moll Flanders*, Ian Watt maintains that "his total subordination of the plot to the pattern of autobiographical memoir is as defiant an assertion of the primacy of individual experience in the novel as Descartes' 'cogito ergo sum' in philosophy."[10] If we replace "individual experience" with "memory" we would be closer to the truth, for the vagaries of memory account for the novel's greatest strength and

---

[9] Ian Watt, *The Rise of the Novel* (Berkeley and Los Angeles: University of California Press, 1967), p. 32.
[10] Watt, *The Rise of the Novel*, p. 15.

weakness. As the eighteenth-century writers tried to ensure duration through affective memory, so the novel maintains continuity by exhibiting the fact that different contents of one's memory at different times belong together, that the recollection of a single, unique event makes it possible to reconstruct one's lifetime. Defoe succeeds best as a novelist in recalling the dozens of unique events in great detail which make up the flesh and bones of *Moll Flanders*. But the novel was a new medium; Defoe felt compelled to better the continuity-of-consciousness through an equal number of expository connective synopses. It is this connective tissue which poses problems for the critic interested in a moral or formal pattern. Although the unique event possesses a neutral valuation, when that event passes through the mind of a morally-concerned novelist it invariably accrues a morality which must be minimally consistent, or at least explicable, throughout the novel. Conceived in haste, this lack of moral continuity is the greatest flaw of *Moll Flanders*. The new importance of memory in man's temporal attitude supports the narrative whose order is determined only in the sequence of recalled events in the lives of the protagonists. It marks a turning point in the history of English literature.

The fact that *Moll Flanders* consists of these brief episodes of action strung along a recollected life is directly related to the increasing awareness of succession as the prime constituent of man's temporal life. Within each episode the new consciousness of time is strikingly evident:

> "The next thing of moment was an attempt at a gentlewoman's gold watch. It happened in a crowd, at a meeting house, where I was in very great danger of being taken, I had full hold of her watch, but giving a great jostle as if someone had thrust me against her, and in the juncture giving the watch a fair pull, I found it would not come, so I let it go that moment, and cried as if I had been killed, that somebody had trod upon my foot and that there was certainly pickpockets there, for somebody or other had given a pull at my watch..."[11]

Daniel Defoe has written a moment by moment account of an actual event. Between each comma lies one of Locke's "sensations," a series of instants linked only by a common perceiver. According to Ian Watt, "Defoe's style reflects the Lockean philosophy in one very significant detail: he is usually content with denoting only the primary qualities of the objects he describes—their solidity, extension, figure, motion and number—especially number."[12] Considering the depth of eternity, Moll confesses that it "represented itself with all its incomprehensible additions, and I had such extended

---

[11] Daniel Defoe, *Moll Flanders* (Boston: Houghton Mifflin Company, 1959), p. 183.
[12] Watt, *The Rise of the Novel*, p. 102.

notions of it that I know not how to express them."[13] While the entire passage is indicative of a moral bankruptcy, this quotation is interesting from a temporal viewpoint for its use of "additions" and "extensions." It is consonant for a period so preoccupied with succession and "enumeration," that eternity should be viewed as an infinite series of additional moments. With the retreat of God from the concept of duration, eternity becomes shallow and successively linear. From this point onwards, eternity never quite possesses the divine immanence which so transfixed men of the Middle Ages and the Renaissance.

3. Richardson's *Pamela*: Accounts of Sentimental Moments with Intense Feeling

In words equally applicable to *Pamela*, Samuel Richardson notes in his Preface to the later *Clarissa*: "All the letters are written while the heart of the writers must be supposed to be wholly engaged in their subjects.... So that they abound, not only with critical situations, but with what may be called instantaneous descriptions and reflections."[14] Again we find the typical eighteenth-century attitude: awareness of the successive moment, but in a rather radical form. Samuel Richardson (1689–1761) is every bit as adept at depicting the moment by moment play of sensations on the mind of the perceiver:

> "She is a broad, squat, pursy fat thing, quite ugly, if anything human can be so called; about forty years old. She has a huge hand, and an arm as thick as my waist, I believe. Her nose is fat and crooked, and her brows grown down over her eyes; a dead, spiteful, grey goggling eye to be sure she has."[15]

But a significant departure from Defoe's practice is also evident in Richardson's Preface; "reflections" may be instantaneous, and portrayed as such; thus *Pamela* far surpasses *Moll Flanders* in depth of feeling. Indeed the epistolary method compels the writer to produce a spontaneous record of a character's emotional reactions; a record which reveals the private orientations of its writer, both towards the addressee and the people discussed. In *Pamela* we find the first instantaneous first-person accounts of the emotional cogito:

> "Oh my exulting heart! How it throbs in my bosom, as if it would reproach me for so lately upbraiding, for giving way to the love of so dear a gentleman! But take care thou art not too

---

[13] Defoe, *Moll Flanders*, p. 251.
[14] Cited by Watt, *The Rise of the Novel*, p. 192.
[15] Samuel Richardson, *Pamela* (London: J. M. Dent & Sons Ltd., 1959), p. 119.

credulous neither, oh fond believer.... will I not acquit thee yet, oh credulous, fluttering, throbbing mischief! That art so ready to believe what thou wishest!"[16]

The rapid fluctuation of emotion from exultation through caution and reluctance presumably corresponds to the passage of these feelings through the minds of Richardson's eighteenth-century feminine audience, and implicitly shares the same time-scheme as consciousness. This new "realism" is partially accomplished by the deft use of punctuation; exclamation marks within sentences, commas, and sentence fragments, which mark the sporadic flux of emotions. Gradually the importance of the moment was becoming imbued and valued by the depth of feeling within the moment. Pamela will proclaim: "May I never survive one moment, that fatal one in which I shall forfeit my innocence," or "Millions of gold will not purchase one happy moment of reflection on a past mis-spent life."

Samuel Richardson's concern for the accurate portrayal of sentiment has unfortunate repercussions for the novel as a whole. While being that most "moral" of novels, *Pamela* is at once both sentimental and obscene. The novelist presents situations; disserts the feeling of his heroine in accordance with the new temporal awareness, and offers the reader a supposedly chaste maiden whose dreams are filled with ideas of rape, but whose waking moments self-righteously resound with necessity of preserving one's honour. This yields the formal danger of the sentimental moment; that in depicting the normal oscillations of feeling, concern for the intensity of feeling overloads the moral balance, and renders it contrary to the novelist's grand design. The ludicrous obscenity of *Pamela* is a direct result of its sentimentality.

4. Sterne's *Tristram Shandy*: Portrayals of the Associative and the Sensational

In the fifty years which separate the publication of *Robinson Crusoe* from that of *Tristram Shandy*, the novel gained a history and the configurations of conventions, both of which were challenged by Laurence Sterne in 1760. The strictness and general acceptance of these conventions is obvious in this anonymous critique of Fielding, written in 1751:

> "A novel may be considered as a dilated comedy; its plot therefore should be uniform, and its narrative unbroken: episode and digression are sparingly, if at all, to be admitted; the early practice of weaving story within story should be avoided...

---

[16] Richardson, *Pamela*, p. 276.

whatever makes a pause in the main business, and keeps the chief characters too long out of sight, must be a defect."[17]

If this dogma were to be rigorously followed by critics today, then *Tristram Shandy* would surely be unique as the most defective novel of the eighteenth century. It is fortunate that such lesser novelists as Conrad, Joyce, Gide, Huxley, and Woolf also chose to follow the defective pattern.

The revolt of Laurence Sterne (1713-1768) against plot, sequence, and causality in favour of structure based on the operative character of consciousness is implicit in the temporal attitudes of the eighteenth century. In the following passage Sterne illustrates his refined awareness of the new temporal thought, and in so doing, provides a most articulate statement of the interactions of succession and duration.

> "For if you will turn your eyes inwards upon your mind, continued my father, and observe attentively, you will perceive, brother, that whilst you and I are talking together, and thinking, and smoking our pipes, or whilst we receive successively ideas in our minds, we know that we do exist, and so estimate the existence, or the continuation of the existence of ourselves, or anything else, commensurate to the succession of any ideas in our minds, or any such other thing coexisting with our thinking—and so according to that preconceived—you puzzle me to death, that in our computations of time, we are so used to minutes, hours, weeks, and months—and of clocks..."[18]

The influence of Locke is perceptible throughout: self-knowledge is built up through association from the primary base of the successive flux of sensations, and this "internal" time is opposed to the purely linear time-line. In fact the opposition between time measured and time perceived forms one of the major themes of the novel, and is the cause of some of Sterne's most hilarious scenes. Generally speaking, two aspects of Lockean theory, the associative and the sensational, account for the external and internal structure of *Tristram Shandy*.

The notion that multiplicity of sensation ensures duration is central to the temporal attitude of the period, and Locke's theory of spontaneous association was the chief philosophic vehicle for the dissemination of that attitude. The seemingly chaotic structure of the novel is merely the non-linear pattern of ideas through the Janus mind of Shandy the novelist; it is unified through the unity of

---

[17] Cited by A. A. Mendilow, *Time and the Novel* (New York: Humanities Press, 1972), p. 162.

[18] Cf. Sterne, cited by Mendilow, *Time and the Novel*, p. 173.

consciousness. However digressional or fragmentary, each section of the novel is the product of a single mind. The novel is organized as reality is perceived and subsequently organized by the mind, as a succession of memories, present experiences, and expectations. For example, in Chapter IV Tristram's birth-date is established and associated with the midwife whose history follows; this in turn leads to Yorick, who first established the midwife, and a description of his horse, which is followed by the general description of the horse, Rosinante, and a discussion of the hobby-horse in general. The progress of the novel is multi-dimensional, whereas that of reality is linear. As A. A. Mendilow points out in his remarkable study, the actions of *Tristram Shandy* occur over a minimum of seventy-five years, and Sterne takes great pains to date each action chronologically against the linear backdrop of history.[19] Thus the apparent disorder of structure is actually reorganized order, organized according to the rules of the mind as expounded by Locke.

The sensational aspect of the moment is not passed over lightly by such a temporally-aware novelist as Sterne. Notice how the following meditation is cut short by the urgency of the moment.

> "Time wastes too fast: every letter I trace tells me with what rapidity Life follows my pen; the days and hours of it, more precious, my dear Jenny! than the rubies about my neck, are flying over our heads like light clouds of a windy day, never to return more—everything presses on—whilst thou art twisting that lock,—see! it grows grey."[20]

Consonant with eighteenth-century experience, the moment that is full of feeling is the most valuable. This doctrine of sentimentality leads to the florid excesses of Uncle Toby and the fly, or the death of Le Fever, again the French influence. Sentiment was central to Shaftesbury's theory that philanthropy produces the highest form of self-satisfaction, which is largely its own reward. If one feels that Sterne is perhaps too philanthropic in empathizing with Le Fever's death, we must remember that the scene ends with the novelist admonishing both himself and the reader for over-indulging.

Sterne's awareness of the reader is but a part of his general awareness of the craft of writing, especially of the temporal limitations of language. Language attempts to organize the flux of consciousness into linear "bits" of information; it is purely successive. The problem is particularly evident when he discusses the double journey to Auxerre, and more so when he calls upon the Gods of

---

[19] Cf. Mendilow, *Time and the Novel*, pp. 188-93.
[20] Laurence Sterne, *Tristram Shandy*, ed. James A. Work (New York: The Odyssey Press, 1940), p. 610.

narration in Chapter XXIII, Volume III. In the controlling consciousness of the novel, three events are simultaneous: the activity in Mrs. Shandy's bed chamber, the anecdotes of Toby's trysts with Widow Wadman, and Toby's campaigns on the bowling green. The successiveness of language compels him to choose one course, and Sterne's awareness of this paradox anticipates the linguistic discoveries of this century.

Given the unique temporal attitudes of the age, a certain pattern is apparent in the progression of the novel from Daniel Defoe to Laurence Sterne. In *Moll Flanders*, Defoe focuses on the sensory detail of the moment; duration is derived from the consistent "I" of memory, but the sentimental aspect of the moment is ignored to the point where the novel lacks a coherence of feeling or morality. In *Pamela*, Richardson is conscious of both aspects, but overemphasizes the sentimental side in his efforts to portray intensity of feeling. This overemphasis results in a paradoxical morality in which the heroine dotes on the act of violation as much as she expresses her honourable disgust. With Sterne, both aspects are fully portrayed in an overall structure which transcends morality by rejecting plot altogether in favour of a structure which is modelled on the reality of mental life, rather than the artificial reality of life filtered through plot. In so doing he created the definitive portrayal of time in the eighteenth century, and a model which novelists such as Leonard Cohen and Samuel Beckett would begin to duplicate a century and a half later.

C. A CASE STUDY OF THE VISUAL IMPERATIVE IN *WATT* AND *BEAUTIFUL LOSERS*

1. The Philosophical Background

The insufficiency of language and the concept of mimesis will be discussed, beginning with a juxtaposition which spans nearly forty years of literary history:

> "Words in modern literature are still being set side by side in the same banal and journalistic fashion as in preceding decades, and the inadequacy of worn-out verbal patterns for our more sensitized nervous systems seems to have struck only a small minority."[1]

> "What I am saying does not mean that there will henceforth be no form in art. It only means that there will be new form, and that this form will be of such a type that it admits the chaos and does not try to say that the chaos is really something else. The form and the chaos remain separate. The latter is not reduced to the former. That is why the form itself becomes a preoccupation, because it exists as a problem separate from the material it accommodates. To find a form that accommodates the mess, that is the task of the artist now."[2]

Samuel Beckett is responsible for the second quotation (from an interview in 1961); Eugene Jolas claims the first, words from the rising radio generation. The combined content of these passages generates two crucial developments in novelistic theory: the refinement of "form" to include the **word** itself as much as the general structure of the total work, and the realization that form is an artifice imposed upon the chaos of reality. Yet this form must somehow act upon the "sensitized nervous system" of the reader, engaged in the act of **reading**. The modern writer becomes acutely self-conscious of form because he no longer talks to an audience, but confronts an unruly mob of prospective participants. He will use any means at his disposal to order their participation towards appreciation of his artistic intent. The form of the novel becomes wholly mimetic in the sense that it imitates the reader's perception and is always an active participation for that reader. This concept of mimesis is not so much a revolution as a rediscovery of the techniques of writers as historically diverse as Rabelais, Sterne, and Joyce. The quotations of Jolas and Beckett are introduced to suggest the intensity of this modern preoccupation with mimesis.

In *Language and Sentence*, George Steiner traces the progress of linguistic entropy from Shakespeare's time to the present. He argues that language "no

---

[1] Cited by Norman Weinstein, *Gertrude Stein and the Literature of the Modern Consciousness* (New York: Frederick Ungar Publishing Co., 1970), p. 1.

[2] Cited by David H. Hesla, *The Shape of Chaos: An Interpretation of the Art of Samuel Beckett* (Minneapolis: The University of Minnesota Press, 1971), pp. 6–7.

longer articulates, or is relevant to, all major modes of action, thought, and sensibility. Large areas of meaning and praxis now belong to such non-verbal languages as mathematics, symbolic logic, and formulas of chemical or electronic relation."[3] This "retreat from the word" received its philosophic underpinnings in the seventeenth century with Descartes' essentially mathematical proof of "truth." The post-Cartesian philosopher uses language in a non-descriptive manner, relying upon the logical consistencies of language to suggest consistencies in non-linguistic reality. The eventual realization that this process of linguistic reference is invalid, that words do not engender responsible apprehensions of truth, and that philosophy ignores the problem of whether reality can ever be spoken of, lies in the genius of Ludwig Wittgenstein. In his aphorisms we discover that we can never hope to know anything about phenomena; we can only know something about the words relating to the phenomena. Beyond the word is nothing. As Beckett's *The Unnameable* states: "It's a question of words, of voices... the words are everywhere, inside me, outside me.... I'm in words, made of words."[4] In a sense then, the modern novelist (especially Beckett and Cohen) does not "retreat from the word" but faces a universe which means nothing but the words he uses. This universe is also private because no one can experience another's experience. The involvement of the reader as participant with its subsequent focus on perception is the novelist's response to this essential privacy; he assumes a common process of perception among his readers.

The primacy of perception generates problems in philosophy and psychology. In the former field, the phenomenological movement of Husserl and Heidegger develops the theory that one cannot distinguish between subject and object as actual entities because the nature of consciousness is projective and directional, always finding itself outside itself. Although such a cursory summation does a great injustice to these brilliant forerunners of existentialism, it becomes rather useful when compared to a current psychological theory—the transactional approach to perception.

> "By perception, then, is meant that part of the transactional process which is an implicit awareness of the probable significance for action of present impingements from the environment, based on the same or similar impingements from the past environment."[5]

---

[3] George Steiner, "The Retreat from the Word," in *Language and Silence* (New York: Atheneum, 1967), p. 24.

[4] Cited by Hugh Kenner, *Samuel Beckett: A Critical Study* (Berkeley and Los Angeles: University of California Press, 1968), p. 56.

[5] Cited by Morse Peckham, *Man's Rage for Chaos: Biology, Behaviour, and the Arts* (New York: Schocken Books, 1967), p. 210.

Each man carries his own "assumptive" orientation through the world, and when that orientation is predominantly verbal, there must be dysjunction. Words are the frozen splinters of reality's flow; they signify only other words, and yet the novelist must use words. The more properly mimetic novelist uses words phenomenologically, with the realization that each reader approaches a page of print through an orientation programmed by past pages, and that for most of us, they are the linear patterns which Eugene Jolas condemns. The mimetic novelist writes against this orientation; he exploits the experience of words as objects in full knowledge that much of the "subject" is made meaningless through words.

"A sentence reproduces in slow motion some mental gesture."[6]

"If we wanted a picture of reality, the sentence itself is such a picture."[7]

"Narration is no longer a line, but a surface on which we isolate a certain number of lines, of points, or of remarkable groupings."[8]

This is the visual imperative, and it may be seen as a paradox of some proportions. On the one hand, as Michel Butor notes: "The sole, but significant, superiority not only of books but of all writing over the means of direct recording which is incomparably more accurate, is the simultaneous exposure to our eyes of what our ears can grasp only sequentially."[9] On the other hand, what is exposed is the inevitable barricade which the word presents between the self and reality. By positing each word as its own reality the mimetic novelist confronts the reader in the act of perceiving, in the paradox of meaning. By joining words visually through dysjunctive syntax, he challenges the reader's orientation, pushing him closer to the truth. In its purest form, the visual imperative is known as concrete poetry. The juggling of perception is a difficult art, and the most successful concrete poems are usually the most minimal in terms of composite units. But "the form must admit the chaos," never become it. The "concrete novel" has not appeared, yet the visual imperative is used by the most innovative writers with telling effects. The nature of the novel is such that the visual imperative always functions within a larger medium; for our purposes the chapter or word-block.

---

[6] Kenner, *Samuel Beckett: A Critical Study*, p. 94.

[7] Ludwig Wittgenstein, *The Blue and Brown Books* (Oxford: Basil Blackwell, 1958), p. 41.

[8] Michel Butor, *Inventory*, ed. Richard Howard (New York: Simon and Schuster, 1968), p. 19.

[9] Butor, *Inventory*, p. 40.

Any word or group of words embodies its own comment both upon itself and upon adjacent words or phrases. It is impossible to determine the maximum "volume" of a word-block which might significantly influence our perception of another word-block. To do justice to an alinear reality, the mimetic novelist abandons the logically connected incidents so characteristic of nineteenth-century narration. In their place, he adopts construction by parataxis which makes possible the creation of myriad relationships among the "blocks" juxtaposed. This process yields a dysjunctive syntax of "chapters" which confronts the reader's usual orientation towards linear meaning, but must (by the very nature of the chapter) ignore the perceptive, which is the basis of the visual imperative. Samuel Beckett's *Watt* and Leonard Cohen's *Beautiful Losers* both employ this technique, although Cohen's is certainly the most exhaustive in this respect.

2. Beckett's *Watt*: The Visual Imperative as an Index of the Nature of Language

Samuel Beckett's *Watt* may be seen as an experiment in linguistic pessimism. Philosophically he is in agreement with Wittgenstein's aphorism: "Whereof one cannot speak, thereof one must be silent."[10] Artistically, the proposition presses him into the most limited of corners described as: "The expression that there is nothing to express, nothing with which to express, nothing from which to express, no desire to express, together with the obligation to express."[11] He will create works without meaning, solely because the meaning of the phenomena is exclusively the meaning of the words which "explain" the phenomenon. The writer's efforts to describe reality invariably dissolve that reality. As Descartes is credited with the historical origin of the belief that language can circumscribe reality, so Beckett creates Watt, the Cartesian hero.

Watt is compelled to form "clear and distinct ideas" of the world so long as they are verbal. He becomes a fictional being, itself composed of words.

> For the only way one can speak of nothing is to speak of it as though it were something. (77)
>
> For to explain had always been to exorcize, for Watt. (78)
>
> But he desired words to be applied to his situation, to Mr. Knott, to the house, ...and in a general way to the conditions of being in which he found himself. (81)

---

[10] Cited by Jacqueline Hoefer, "Watt," in *Samuel Beckett: A Collection of Critical Essays*, ed. Martin Esslin (Englewood Cliffs: Prentice-Hall Inc., 1965), p. 75.

[11] Cited by Kenner, *Samuel Beckett: A Critical Study*, p. 30.

And Watt's need of semantic succour was at times so great that he would set to trying names on things, and on himself, almost as a woman hats. (83)

But he had turned, little by little a disturbance into words, he had made a pillow of old words, for a head.[12]

With the creation of Mr. Knott's establishment, Beckett posits a reality which refuses to be named; the events remain meaningless for Watt and the reader. The disharmony of Knott's piano is extended by the Galls to include Mr. Knott, the centre of non-meaning. Watt painfully realizes "that a thing that was nothing had happened, with the utmost formal distinctness, and that it continued to happen..." (76). The Cartesian hero's most telling encounter with nothingness in the verbal sense, involves the pot that refuses to be named; it must always stand beyond the barrier of linguistic definition. Eventually everything in Knott's house moves beyond the range of words, and Watt, the man of words, moves to an insane asylum.

As long as Watt is capable of locating an event in a linear series, he is reasonably comfortable. The splintering movement of language is also linear, through cause and effect, and subject-verb-object. In the mimetic novel which imitates the reader's perception, the technique often becomes the content and Watt's breakdown is expressed in a breakdown of the line which connects past to present and future. Thus the major word-blocks of Watt are not chronologically ordered. The parallel breakdown of the hero and language underlies Beckett's simplest use of the visual imperative.

The reader's eye travels across the page in a fashion which would please Watt. The sinuous considerations involving the selection of Knott's table-dogs meander through ten pages before the "imperative" flashes from the page. Beckett presents a diagram, a chart explaining the number of objections to each possible solution. The object on the page is simultaneously a condensation of Watt's mental configuration (a list being a vertical series,) and the point at which the reader perceives the essential linearity of what has passed before.

The participation of the reader is more pronounced in Section Three where Sam offers examples of Watt's manner of speech while at the asylum. A change to italic type-face is the visual cue; a reversal of syntax, the content. Watt's verbal obsession is obvious when one considers the reasoning behind the syntax, that he is merely rearranging the order of the series. Not only does the technique exactly mirror the content, the reader's response to the new order must be one of immediate surprise, even discomfort if he tries to extract meaning from the

---

[12] Samuel Beckett, *Watt* (New York: Grove Press Inc., 1959), p. 117. All further references to this edition will appear in the text as a page number within parentheses.

series. Is this not the same mental attitude of Watt himself? In other words the visual imperative in this and succeeding examples operates to fuse the reader's perception with the character's mind, and as Watt is language the perception becomes one of the insufficiency of language.

Beckett's most complex use of the visual imperative involves ironic distance between the reader and the work. The effect of this encounter is to impel the reader against the ostensible reason for the presentation of the page. The celebrated incident in which the reader sees the frogs' voices is a prime example. The ostensible reason for the display is to show how Watt heard a threnody of frogs as harmonic progression. But the reader perceives the page simultaneously and thereafter his gaze does not travel "across" the silence, but rather diagonally down from left to right through the angle of major signs. The net effect is the perception of silence punctuated at intervals by amphibian voices, but the intervals are irregular, the object-in-itself is "a form which admits chaos," not Watt's supposed perception of harmony.

A similar effect is noted with regard to the musical "picture" heard by Watt while in the ditch. The initial effect must be of the "one-ness" of the double-page spread; in other words, its simultaneity, and the reader is seemingly plunged from language into something approaching music. But a musical score is as linear and fragmented as language, and harmonies ensue only upon its being played and heard. No reader perceives the musical and vocal harmony presented by the vertical "harmony" of "fifty-fifty-fiffee-hem!" (34). The presentation of Watt's ideal harmony is ironically undercut by the reader's perception of that harmony as a linguistic series. This use of the visual imperative is consonant with the doomed piano of Knott's house and might be seen as a foreshadowing of the "Knott-linear" chaos which is the undoing of the Cartesian hero.

In Watt, the least visually spectacular use of the imperative is also the most common; the numerous repetition of words, and the lengthy permutations involving but a few words. Of this, the smallest example will refresh the reader's memory:

> "...and mothers' fathers' and fathers' mothers' fathers' and mothers' fathers' mothers' and fathers' mothers' mothers' ...and fathers' fathers' fathers' and mothers' mothers' mothers'." (46)

This segment of a virtually endless series illustrates that repetition of words to the point where they lose their meaning, and become mere inkspots on the page. The reader perceives the series developing, perhaps a glance confirms the continued appearance of identical terms, and removes himself from the process of meaning-extraction. In each case the visual imperative provides a retraction from the word as representative of reality, and reinforces the falseness of Watt's verbal reality. That this should be the most pervasive stylistic device of the novel, returns us to Beckett's aesthetic pessimism, the abyss which separates the world from the word.

In *Watt*, that arrangement of terms on a page which we have called the visual imperative invariably returns to, and compels the reader to, the realization of the absurdity of language. Words are arranged that their apprehension undermines the reader's orientation towards words. Frank Kermode provides us with a terse Beckettian summary. "There are no circles in reality, and no novels."[13]

3. Cohen's *Beautiful Losers*: The Visual Imperative as an Index of the Nature of Consciousness

When we turn from *Watt* to Leonard Cohen's *Beautiful Losers* we move from a novel of language as futility, to a novel of language as transcendable obstacle. Unfortunately it is also a move from a world-acclaimed Nobel prizewinner (with the company of accomplished critics) to a little understood Canadian writer whose critics are few and divided. For this reason, more attention must be paid to such fundamental questions as theme and meaning; unlike *Watt*, *Beautiful Losers* has never received the critical merit it deserves.

In another essay in *Language and Silence*, "The Pythagorean Genre," George Steiner writes that "wherever literary structure strives toward new potentialities, wherever the old categories are challenged by genuine compulsion, the writer will reach out to one of the other principal grammars of human perception—art, music, or more recently, mathematics."[14] Regrettably, he has ignored that most recent fertile field of human perception, the cinema. Much of *Beautiful Losers* remains inaccessible until its structural debt to cinema is realized.

The most important aspect of the movies which modern writers draw upon is the technique of montage. The cinematic montage may either express harmony and stillness, or cacophony with erratic movement in any direction. In *Beautiful Losers* the first technique is evident in chapters two through five when various aspects of the Narrator's obsession with Catherine are revealed. The second type of montage (romantic juxtaposition) is developed in the format suggested by Sergei Eisenstein: "By collision. By the conflict of two pieces in opposition to each other. By conflict. By collision."[15] The process leads to a radical "syntax" wherein each word-block relates across a temporal pit of unknown

---

[13] Frank Kermode, *The Sense of an Ending* (New York: Oxford University Press, 1967), p. 149.

[14] Steiner, *Language and Silence*, p. 87.

[15] Cited by Sharon Spenser, *Space, Time and Structure in the Modern Novel* (New York: New York University Press, 1971), p. 116.

dimensions as in the mosaic. For example, section thirty presents a frenetic interpretation of a pop record which becomes a humorous yet tragic ritual. These energetic impressions of rock-and-roll sensuality collide with the next section, a low-keyed enigmatic confession. This is then played against a return to the impersonal facts of seventeenth-century history in the following section.

These myriad juxtapositions also reflect a mobile concurrency of time in which flashbacks and flash-forwards may occur at random. By interposing consecutive lines from Books One and Two, an interesting time-warp is revealed:

> PERHAPS THE TREEHOUSE WHERE YOU SUFFER IS THE
> HUT OF OSCOTARACH.
> IS THIS TREEHOUSE THE HUT OF OSCOTARACH? (168)
> ASK YOURSELF. WAS I YOUR OSCOTARACH? (232)
> WERE YOU THE HEAD IN PIECES,... ? (168)
> THE SURGERY IS DEEP IN PROGRESS (232)
> IS THE SURGERY DEEP IN PROGRESS? (168)[16]

Moreover, when we focus on the action within the montage, other cinematic techniques become evident: F.'s long letter concludes as a scenario complete with shot directions for invisible cameramen. More subtly, Cohen uses the camera eye to present an event from many different perspectives within a short length of prose. A combination of perspectives brought into simultaneous play expands the potential expressiveness of a work. With Cohen the technique is used to create a saturation of significant actions which annihilates ordinary narration by the sheer weight of impressions.

Such use of movie techniques points to a fusion of method and metaphor, the idea of technique as content which is so implicit in mimesis. Thus thematic metaphor is narrative content, and the movies are woven into the novel as both a presentation of a higher state of consciousness and as a radical structural technique. Guidelines for the attainment of this state are offered in F.'s epigrams and these become "syntactic" variations. Early in the novel F. advises that one must "Connect nothing.... Place things side by side on your arborite table, if you must, but connect nothing" (21). There is a striking resemblance between this message and the colliding shots of the montage which in the novel are placed "side by side" as chapters and juxtaposed scenes.

A similar fusion is evident in Cohen's cinematic treatment of time. As F. lyrically proclaims: "Listen, my friend, listen to the present, the right now, it's

---

[16] Leonard Cohen, *Beautiful Losers* (New York: The Viking Press, 1967), p. 23. All further references to this edition will appear in the text as a page number within parentheses.

all around us.... Empty your memory and listen to the fire around you" (14). Compare the above passage to Gertrude Stein's comment on the cinema: "By a continuously moving picture of any one there is no memory of any other thing and there is that thing existing, it is in a way if you like one portrait of anything not a number of them."[17] We have already noted how the suggestion of cinematic presentness is achieved by the accretion of impressions, but on a larger scale the interface between the presence of consciousness and the presence of the movies as syntax is implied in the alinear arrangement of chapters. As distinctions between past, present, and future disappear without concrete signposts, the novel emerges as a single, almost transparent event in which the various situations posed within the novel may be seen as concentric ripples radiating from the depth of the event, but not its succession. The event of *Beautiful Losers* is the enlightenment of the fused "beautiful losers" in that moment when "the future streams through... going both ways. That is the beautiful waist of the hourglass!" (305). Leonard Cohen's novel must be seen as a book which admits succession only in its consecutively-numbered pages.

Late in the novel, F. proclaims the credo of the mimetic novelist. "Watch the words, watch **how it happens**" (235). *Beautiful Losers* is continually showing us how it happens. Book One provides a precise chart of the Narrator's psychic progress as reflected in style and the visual imperative. The use of the list or catalogue in section twelve carries the reader into the realm of useless details which is in turn linked to the bondage of language and history. As F. states: "Of all the laws which bind us to the past, the names of things are the most severe.... Science begins in coarse naming..." (51). The narrator "live(s) in a world of names" (22) and has "lost (him)self in particulars" (34). In section ten the visual imperative operates through repetition to underscore this preoccupation. We find more than two pages of detail piled upon detail; each is preceded by the statement "I wanted." The effect on the reader is a mental imprint of the Narrator's essential "selfishness" which is the greatest barrier to liberation. The connection between details and the self is most obvious when the Narrator declares: "I am the sealed, dead, impervious museum of my appetite" (50). At this point he compares himself to a repository for the details of history and then links this to the "appetites" of desire. The necessity of appetite for health is self-evident and the corollary, that the mind's preoccupation with the "eating" of details is as strong a bond, returns the novel to its central theme of the liberation of consciousness.

The burden of the catalogue as a list of details strung along a vertical line is implicit in constipation, or being full of one's self. "I thought Edith would rest in a catalogue.... I forgot about constipation! Constipation didn't let me forget. Constipation ever since I compiled the list" (47). As the body eats food,

---

[17] Spenser, *Space, Time and Structure in the Modern Novel*, p. 118.

so the mind consumes details. When the mind is preoccupied with the data of details (as in the Narrator's role as historian), they become rigid in the bowels of memory as obsessions perhaps, and the mind is unable to free itself into an experience of the pure present.

Against the many pages crowded with details of the past, Cohen presents the magical view of the native Indian. On a nearly blank page (the blankness corresponding to silence and the emptiness of the self), the reader's eye is drawn to the Indian's chant which bisects the page while drawing that space together. "I change... I am the same... I change... I am the same.... Every change was a return and every return was a change" (166). The magical cyclic time of the Indian contrasts sharply with the linear Christian time-line which science has dissected into names and history.

As appetite reaches out for other things besides food we would expect some aspects of sexual delight to be connected with the psychic burden of history. The exploration of this link is the chief concern of section twenty-five, a chapter which uses the visual imperative in an unpunctuated, staccato monologue of memory and fancy. The syntactic innovations of this word-block propel the reader into the intense psychic discontinuity of sexual stimulation. But masturbation is another instance of preoccupation with self-detail, and a denial of one of F.'s most important teachings, the belief in the pan-orgasmic body:" All parts of the body are erotogenic.... All flesh can come" (40). In denying F., the Narrator is closing one door to enlightenment; as a result he is locked within himself and the monologue is fraught with self-despair: "Thy lonely husband... thy lonely husband... thy dark lonely husband." (84)

As Cohen uses radical syntax within the visual imperative to carry the reader past those psychic barriers which inhibit liberation, we would also expect such usage to underscore the higher states of consciousness as the Narrator progresses spiritually. In section twenty-nine, the comic book scene, we are presented with peculiarly modern icons of the quest theme, specifically, F.'s idea that Charles Axis should be followed because he is the one man who contains all of mankind's frustrations and all of its fantasies. Continuing the metaphor of movies as the highest level of consciousness, we find a cinematic scenario in which the seven panels of the strip correspond to separate shots of the montage. Each panel is narrated to follow the reader's perception of the comic strip: broad scene-recognition followed by increasing detail, and various typefaces strike the eye with a force determined by their relative size. With this radical approach the usual confessional style disappears as the Narrator significantly forgets his self-consciousness and becomes, as it were, one of the theatre audience. The use of the visual imperative assumes greater importance when we realize that this flash-back scene marks a turning point in the lives of F. and the Narrator. The former begins daily exercise, while the Narrator drifts into the "sin of pride" and a further preoccupation with self.

In a significant juxtaposition, the next chapter, the radio scene, presents another stage of psychic growth through the unique imperatives of ritualistic drama. Because the radio and Greek tragedy tend to unite "psyche and society into a single echo chamber,"[18] this scene is marked by another disappearance of the Narrator's "I" in favour of a more detached point of view. For the first time, he moves outside himself into the "cold ordinary world."

In chapter forty-four, the Narrator discovers the nature of the relationship between the deceased "beautiful losers." This stage is marked by the presentation of facsimiles of advertisements which place the transcendent methods "side by side." At this point the Narrator disappears altogether; the reader faces the object of bewilderment exactly as it appeared to the novel's narrator. His questions are the reader's questions. The novel imitates the reader's perception to the point where the reader becomes a character in the novel, and the explanation which follows assumes added importance.

The chaotic concluding sections of Book One are usually interpreted as the Narrator's furthest descent into madness. But if this view is accepted there is much less justification for his eventual enlightenment in the Epilogue, and complaints arise that *Beautiful Losers* lacks synthesizing power.[19] If, however, the correspondence between the visual imperative and super-verbal consciousness is maintained, the last two chapters point directly to the achievement of sainthood. Thus although section fifty-one is composed of a list of details which is reminiscent of chapter ten, a new syntax in which the subject "I" has disappeared is evident. Moreover the active verb "wanted" has been replaced by a more passive "accepted," and this stance of acceptance is prerequisite to revelatory religious experience.

As psychic advancement continues through the final chapter of Book One, even the implicit "I" of the previous section disappears and is replaced by the new "I", which is the emptiness of the phrase-book "I" in which any reader may place himself. The visual imperative presents a consciousness equidistant from the Narrator and the reader, yet it is perceived in the same manner by both, as language imposed upon a "block" of silence. The phrase-book exists apart from the Narrator as a prayer into which one pours oneself once the mind has been cleansed of details. It is this connection to which F. alludes when he states that "Prayer is translation" (71). The content of the prayer suggests those pagan rites in which the initiate prepares for the final voyage of the soul by improving his appearance, bidding farewell to friends and relatives, and gathering those articles which he will find useful in the afterlife. When the

---

[18] Marshall McLuhan, *Understanding Media: The Extensions of Man* (New York: McCraw-Hill Book Co., 1965), p. 299.

[19] Cf. George Woodcock, "The Song of the Sirens: Notes on Leonard Cohen," in *Odysseus Ever Returning* (Toronto: McClelland and Stewart, 1970), p. 108.

initiate is ready for final meditations, the Narrator appears briefly to offer calm reassurance: "O God, I grow silent as I hear myself begin to pray" (180).

The radical syntax which underscores key transitions in Book One contrasts sharply with that of Book Two which, as Michael Ondaatje notes, "shows a remarkable change in style, a calm that is gracious after the diatribes and uncertainties of the Narrator."[20] While Ondaatje is quite correct in this observation, he does not delve deep enough into the significance of this shift. F. wears his style in a symbolic fashion. F. alludes to this notion near the beginning of his long letter:

> "This letter is written in the old language, and it has caused me no little discomfort to recall obsolete usages. I've had to stretch my mind back into areas bordered with barbed wire, from which I spent a lifetime removing myself." (193)

As befits his position as master or teacher, F. has escaped or more exactly transcended the prison of linear syntax and now returns to the old style, only to offer instruction for passing the boundaries. To reveal the magic in reality, he utilizes two minor aspects of the mimetic style: the aphorism and illogical imagery. Through juxtaposition the epigram dilutes the causality inherent in language and at the same time suggests the matured awareness of the author who sees a unity where none was present before. For the reader it becomes a creative act of perception which "by connecting previously unrelated dimensions of experience enables him to attain to a higher level of mental evolution. It is an act of liberation—the defeat of habit by originality."[21] Again, mimesis challenges the reader's linear orientation. With the second syntactic variation, F. adheres to his dictum of "connect nothing" and places impressions side by side on his arborite-table with the result of imagery which is both illogical and brilliant:

> "Mary is bouncing her bum like a piggy-bank which is withholding a gold coin.... Delicious soup stews my hand. Viscous gupers shower my wrist. Magnetic rain tests my Bulova. She jiggles for position, then drops over my fist like a gorilla net." (196)

Although the structure is linear, the associations deny logical connections. The reader is forced to close the metaphors with his own imagination; he must "watch the words" and "make it happen," a lyric approximation of mimesis.

---

[20] Michael Ondaatje, *Leonard Cohen* (Toronto: McClelland & Stewart. 1970), p. 47.
[21] Arthur Koestler, *The Act of Creation* (London: Pan Books Ltd., 1969), p. 98.

As Book Two concludes with the "history" of Catherine's asceticism and enlightenment, the presence of liberation demands a return to the alinear structure of word-blocks. As F. "tells" the story of his own trials and enlightened escape, the visual imperative returns as an index of liberated consciousness, in the guise of a cinematic scenario. The reader experiences the implied replacement of language by instantaneous electronic media: radio and cinema.

With Book Three Leonard Cohen sidesteps the problem of presenting death (as loss of self) in the confessional first person narrative by resorting to the omniscient third person. As the "I" of self-consciousness dissolves, possibilities for a syntactic model disappear as well; disjunctive syntax is removed from the field of consciousness and into action where it is used to enhance the vitality of the reported event. Thus the visual imperative appears in the reproduction of pin-ball instructions in a contrasting typeface to present the event by bypassing the filter of narrative consciousness. Again presentness is achieved by the piling of image upon image to suggest multi-faceted perspectives. With the fusion of the two main characters in the Clear Cinema Light of the Second Chance (or Bardo in Tibetan religious doctrine), the contents of consciousness move out into the streets and "the Newsreel escapes into the Feature" (283). As consciousness transcends self, cinematic syntax captures the electronic environment which surrounds and pervades the event.

Any mimetic novel presents problems for criticism which expects linear patterns of coherence and expectation to be reinforced. This problem of coherency is the chief stumbling block of *Beautiful Losers*. Sandra Djwa responds to the novel's dysjunctiveness by focusing on the details of the "antihero's" disintegration, but fails to see any "direction suggested for the movement beyond recorded disintegration."[22] Sharing a similar viewpoint, George Woodcock believes that "these experiments are interesting; little else. For they are not integrated into a functioning unity."[23] By concentrating on the "dysjunctive" factors of montage and the visual imperative, supposedly the weakest links of the linear chain, a unity has been shown in the fusion of technique and content which distinguishes the mimetic novel. In *Beautiful Losers* we noted that the visual imperative was used to propel the reader through stages of the Narrator's psychic bondage, especially the tyranny of language and history. This being established, the imperative underscores climactic moments in his quest for liberation. Syntactic variations are played against each other to suggest the dynamics of the see-saw process of the mind striving for illumination. This approach is balanced by that of Book Two where F., as teacher and

---

[22] Sandra Djwa, "Leonard Cohen, Black Romantic," in *Canadian Literature*, 34, (1967), p. 42.

[23] Woodcock, *Odysseus Ever Returning*, p. 109.

possessor of a more evolved consciousness, "puts on" syntax for didactic purposes and thus his psyche is delineated without the process of his evolution being documented. The first book then uses what might be termed (from the thematic point of view) a dynamic imperative, whereas in the second it is essentially static.

The importance of theme cannot be overemphasized. Because both novels share the concept of mimesis with its focus on the reader's perception, and because the visual imperative is an essentially neutral device within the technique of mimesis, the shape of the imperative must mirror its thematic content. In *Watt*, we face a narrative experiment which exploits the failure of language (as a function of reason and logic) as a means of apprehending reality. The various instances of imperative are designed to represent this failure in the reader's mind. Watt as the Cartesian hero is language itself and the visual imperative becomes primarily an index of the nature of language, and only secondarily an index of consciousness. However in *Beautiful Losers* (a novel which admits the transcendence of language by assuming the cinematic metaphor) the nature of consciousness is primary and the barrier of language, secondary. That the visual imperative can accommodate both viewpoints is a testament to the genius of both novelists, and that such a statement can be made in the first place, a testament to the utility of this concept.

D. SUMMARY: THE MYTHOPOEIC IMPOTENCE OF THE MODERN WESTERN MAN

A myth composed by a modern artist, even one of Joyce's stature, is different from ancient myth. Certainly a modern myth is more conscious, more private, and its use of irony is quite absent in the original models. These three characteristics, especially the privacy of a "monomyth" within what is surely the least accessible of modern novels, lead one to conclude that Joyce's novel is not so much a myth as an artistic attempt to rehabilitate the cyclic myth for modern man. That the attempt is a unique achievement of monumental genius is not disputed.

Let us reconsider how the modern mentality defies both the esoteric "Great Wheel" myth of Yeats and the pan-cultural Viconian myth of Joyce's *Finnegans Wake*. In essence, twentieth-century Western man has experienced, by adulthood, an almost complete displacement of the modes of consciousness necessary for mythology to exist as an explanation of reality. Henry Murray's brilliant essay, "The Possible Nature of a Mythology To Come," outlines the extent of this transformation from archaic man's "child-like" mythologizing to the extreme rationalism and consequent alienation of modern man. Four aspects of displacement are indicated:

1) Emotional identification with nature and projections into space are displaced by cognitive detachment, allowing dissections of the environment into concepts of material particles and energy;
2) A progression from non-verbal images and feelings to emotive diction, followed by the theoretical diction of abstract concepts, and finally to symbols ultimately dissociated from images and feelings;
3) An initial receptivity to visions, sensory impressions, and authoritative statements changes with cultural evolution and aging, to a state of suspension of judgement in the absence of evident proof;
4) An ever higher standard as to what constitutes sufficient basis or evidence for a statement.[1]

These then are the psychological roots of our modern incapacity to embrace a mythopoeic perspective in the twentieth century.

Naturally, the extent to which literature transcends these limitations is the extent of its capacity to present myths; and as we have seen in Yeats and Joyce, the world of myth lies very close to the centre of their literary preoccupations. Lamenting the separation of fact and value, T. S. Eliot named it "the dissociation of sensibility," a cultural phenomenon which implies the impossibility of art because we have come to accept fact and empirical evidence as the total reality. By attempting to rehabilitate the cyclic myth as that which is immanent behind

---

[1] Adapted from Henry A. Murray, *Myth and Mythmaking* (New York: G. Braziller, 1960), p. 314.

empirical reality, Yeats and Joyce attempt to restore to art its cosmogonic function of explaining the true nature of reality. On the other hand, the shift of temporal thought is evident from the Renaissance to the modern time. It had a great imprint on the rise of the English novel, as the eighteenth-century writer experienced life as a successive series of moments connected by a duration based on personal memory and multiplicity of sensation. As the successive nature of language limits it to organize the flux of consciousness and as language fails to be a means of apprehending reality, the modern novelist has thus a private universe to surpass. But before comparing the origins and transformations of the cyclic myths in Chinese and Western cultures, it is imperative for us to demonstrate, through a study of the philosophical message in the works of Ibsen, Chekhov, Lu Hsün, Beckett and Pinter, the phenomenon of cultural convergence in literature as a parameter of the "time spirit" of an age, in order to reaffirm, under the same cultural phenomenon, the supposition of the cyclic myth as a universal archetype in both cultures.

IV. THE CULTURAL CONVERGENCE AND THE AGE'S "TIME SPIRIT"

A. A CASE STUDY OF THE PHILOSOPHICAL MESSAGE IN THE WORKS OF IBSEN, CHEKHOV, LU HSÜN, BECKETT, AND PINTER

1. The Cultural Convergence and the Age's "Time Spirit"

At first glance, a discussion of the interactions between literature and philosophy would seem to be one of the original inhabitants of Pandora's box; great are the dangers of fabricating intellectual constructions which appear habitable only from a distance. No one would hesitate in labelling Georg Hegel's "*Philosophy of History*" as philosophy, or Henry Fielding's *Tom Jones* as literature. But can we be as certain with Friedrich Nietzsche's *Thus Spoke Zarathustra* or Alexander Pope's *Essay on Man* ? The problems begin almost before the lid is lifted.

The philosophical approach to literature is by definition concerned with ideas; specifically those of metaphysics and eschatology. The term "idea" refers to our more reflective or thoughtful consciousness as opposed to the immediacies of sensuous or emotional experience. Both arts are dependent upon thought and imagination, and we might maintain that the introspective, almost psychological way of literary insight is precisely that method of "inlooking" by which the philosopher, in his spirit of inquiry, descends below the level of mere appearances to what we could call "truth" or "knowledge." A similar point is made by Theodore Hunt: "Philosophy owes to literature the imaginative and emotive impulse by which philosophic truth is set forth."[1] The thinker is concerned with implications and seeks to preserve a more or less rigorous logical consistency in the presentation of his idea, whereas the imaginative writer is more apt to make clear how an idea affects life within the free, often illogical, boundaries of the work of art. He presents a mental product for realization; the philosopher, a mental problem for inspection and solution.

In *Theory of Literature*, René Wellek and Austin Warren devote a chapter to the various philosophical approaches to literature.[2] These range from Lovejoy's "History of Ideas" where literature is used only as document and illustration in tracing a chronology of philosophic fragments or "unit ideas" to that most extravagant school of "*Geistesgeschichte*" (Intellectual History). In this view literature is seen as a parameter of each age's "time spirit," providing that one can conceive of a period of history as an integrated, mutually-enforcing unit. The authors find merits in each approach but deplore the almost necessary

---

[1] Theodore W. Hunt, *Literature: Its Principles and Problems* (New York: Funk & Wagnall, 1906), p. 72.

[2] Cf. Rene Wellek and Austin Warren, *Theory of Literature* (New York: Harcourt, Brace & Co., 1949), pp. 107–23.

trend towards intellectual reduction of a work of art to a rung in some critic's systematic ladder. This conclusion should be self-evident to anyone with a sincere regard towards the autonomy of literature as art: "Philosophy, ideological content, in its proper context, seems to enhance artistic value because it corroborates several important artistic values: those of complexity and coherence."[3]

Insofar as literature and philosophy should be viewed as an on-going dialectic of ideas, one approach (at least in its initial conception) seems especially valid. Rudolf Unger argues that "literature is not philosophical knowledge translated into imagery and verse, but that literature expresses a general attitude toward life, that poets usually answer, unsystematically, questions which are also themes of philosophy but that the poetic mode of answering differs in different ages and situations."[4] This phenomenon, that different individuals in the same culture frequently arrive at the same solution to a problem, but quite independently of each other, is known as cultural convergence. Authors of different countries or varying ages may perceive the problem independently; yet independently they would arrive at the same solution. Our next task shall be to pose the philosophic "problems" in which the plays of Henrik Ibsen (1828–1906), Anton Chekhov (1860–1904), Samuel Beckett, and Harold Pinter, and the novel of Lu Hsün (1881–1936) may be seen as unsystematical "answers."

According to Robert Brustein, the drama of the Western world "describes a trajectory which arches from belief to uncertainty to unbelief, always developing in the direction of great scepticism towards temporal and spiritual laws."[5] We might add that this "arch" metaphor probably describes the direction of European philosophy with greater similitude than it does modern drama. Besides this, the comparison between Lu Hsün and the other four authors in this section is made possible by the fact that the May-Fourth Movement in China greatly increased the number of literary translations, in which Russian literature and German literature were most popular. However, the works of Ibsen and Chekhov were more widely read. And it is the underlying spirit of social criticism, analogous to that of Ibsenism, which makes possible a comparison of Lu Hsün's *The Biography of Ah Q* with plays such as *The Wild Duck*, *The Seagull*, *Happy Days*, and *The Homecoming*.[6]

---

[3] Wellek and Warren, *Theory of Literature*, p. 122.

[4] Wellek and Warren, *Theory of Literature*, p. 113.

[5] Robert Brustein, *The Theatre of Revolt: An Approach to Modern Drama* (New York: Atlantic Press Monthly Book, 1964), p. 5.

[6] For the influence of Ibsenism and Russian plays upon the rise of modern Chinese drama, cf. Edward M. Gunn, ed., *Twentieth-Century Chinese Drama: An Anthology* (Bloomington: Indiana University Press, 1983), pp. vii–xx.

As the nineteenth century progressed, the transcendent frame of reference for truth, so convincingly established by Immanuel Kant (1724–1804) in the late eighteenth century, decayed into a variety of more "scientific," ever more ingenious hypothetical systems culminating in the ponderous mechanisms of Hegel. When Ibsen left Norway in 1864 to begin his great epic dramas in Rome, scientific naturalism was the world view of civilized Europe: Karl Marx (1818–1883) had just completed his *Critique of Political Economy* which traced the development of culture solely on the basis of class interests, and Charles Darwin (1809–1882) published his *Origin of the Species* five years earlier. If the well-educated man of the age was sceptical about God, he could not doubt his own rationalism. Friedrich Nietzsche (1844–1900) was the first to realize that if the Philistine's life was comfortable, it was also a most outrageous lie.

Modern drama rode in on the second wave of Romanticism; not the cheerful optimism of Rousseau, but rather the dark fury of Nietzsche, with his radical demands for a total transformation of man's spiritual life. And Nietzsche remains the most seminal philosophic influence on modern drama, the intellect against which almost every modern dramatist must measure his own. As products of the same all-important cultural forces, the dramatist need not have studied Nietzsche's philosophy to find himself grappling with typically Nietzschean problems.

If, in ignoring the facts of life as "lived," the rational philosophers could be termed dishonest; perhaps then the first requirement of intellectual honesty is to accept life exactly as it is to the exclusion of any want beyond it, with all its physical drives and dark impulses, as well as the spiritual power of man's intellect and creative forces. This process of life is not, as Hegel thought, a dialogue of the mind with itself, but the true dialectic that lies in the ever-changing encounter between the mind's drive for order and the chaos of reality in a continuous restructuring of orientations, both cultural and psychological. In that dialectic process between reality and thought lie identity, order, meaning, and therefore value; these are constantly lost as concepts dull and break on reality's contradictions, and constantly renewed as the mind forges new instruments for these encounters. The world is nothingness; we awake from nothingness to struggle with nothingness, and in that struggle we forge, and continually reforge, our identities. The world is without order, without meaning, without value. Human identity has no ground. The world is nothing, but in emerging from that nothingness and encountering it, we create our being.

2. Ibsen's *The Wild Duck*

While Nietzsche insists that life must be lived in this "superman" way, the dramatist must portray life as it is and put the dream to test; how can illusion be made reality? The dramatist seeks to exalt the ideal, yet remains imprisoned in the real. He would vindicate the self, yet he must also examine the claim of others. He would sing of freedom, yet he must cope with the tedious, conditioned world. Thus modern drama is extremely self-conscious: the Romantic who would destroy all boundaries is also a Classicist, accepting limitations in life and art.

> "He is an absolute anarchist, wants to make a tabula rasa, put a torpedo under the ark; mankind must begin at the beginning of the world—and begin with the individual.... The great task of our time is to blow up all existing institutions—to destroy."[7]

So wrote an outraged Philistine following a conversation with not Nietzsche, but Ibsen; this was in 1883, one year before Ibsen wrote *The Wild Duck*. It is a measure of the conflict in Ibsen regarding the transformation of the ideal to the real that *The Wild Duck* should result not in the fiery vortex of *Zarathustra*, but a puzzling play of illusion, accommodations and adaptations in which the injured and the deprived create forms of innocence and order as stratagems to avoid confronting the self. Ibsen is interested in the avoidance of truth as a valid response to life, the dream must be tested in an actual situation. The interplay between truth and illusion in *The Wild Duck* is best exemplified in the pervasive metaphor of sight, blindness, and photography.

Hjalmar is the reluctant retoucher of photographs, the sentimental faker of reality; Gina the literal realist, the one who actually takes the pictures; Hedvig, their daughter, lives in both worlds and, through her preference for drawings, centres into the imaginative world of childhood and art.

The subjects of the photographs are happily-married couples; the pictures are of an ideal order. A pattern develops in which the harmony of appearances is systematically shattered by revelations of biological heredity (Ibsen was no doubt familiar with Darwin) and sexual transgressions. In other words those darker sides of life which Nietzsche discovered and incorporated in his view of freedom rise up and grant not freedom but chaos and death.

It is the function of illusion to provide the personality with fixed patterns of value, which are nothing but orientative patterns in the mind that guarantee a certain amount of meaningful continuity. This is not free; it must be transcended before the ideal and the real may merge. Enter the rebel, the revealer; Gregers is a transcendentalist with a sick conscience. His "claim to the ideal" is futile and destructive in that it confronts the victims with the truth of their victimization and provides them with no sustaining notion of order to render that truth acceptable or even bearable. In fact Ibsen, through Dr. Relling, suggests that Gregers use Hjalmar in substitution for his own identity in motives that sound

---

[7] Brustein, *The Theatre of Revolt: An Approach to Modern Drama*, p. 38.

more like revenge than salvation. As the thirteenth man at the table, he is more like Judas than Christ.

On the other hand, Ibsen portrays the world of illusion as exemplified in the loft as a degeneration; the various accommodations which guarantee survival in the hostile real world can be seen as a kind of death. The problem hangs in a balance between the truth which kills and the illusions which amount to a living death.

If Gregers' "claim to the ideal" is examined closely, however, it begins to sound suspiciously like the old religious impulse which substitutes a loftier illusion for the common, everyday evasions of chaotic reality. Gregers posits an absolute truth to people who are incapable of transforming such a truth into an actual incentive for their lives, and as such, it becomes inverted as Gregers' own escape from reality.

The balance seems to be in Ibsen's implied belief that truth is nothing absolute but in constant flux. By posing the extremes of the transcendental and the fantasies of common people against each other we may arrive at the perception that the truth is never a possession. It is a constant effort to find the appropriate response to every given situation which demands a decision, and that truth, once it is generalized, is already disintegrating.

Thus, however obliquely, Ibsen returns to the existential stance of freedom, although by now the superman has lost his fiery sparks and is replaced by the man who endures.

3. Chekhov's *The Seagull*

Unlike Ibsen, Chekhov was not a man of ideas and definitely not a deep thinker in the sense that Dostoevsky reveals himself. Chekhov himself admits: "All I wanted to say was 'Have a look at yourselves and see how bad and dreary your lives are'."[8] Certainly if all that Chekhov's "philosophical message" amounted to was a revelation of the ultimate hollowness of life, we might agree with F. L. Lucas and say that *The Seagull* is "full of highly sensitive but futile persons, living in a dark, stuffy little jungle, mutually devouring and being devoured."[9] But there is more.

Chekhov's attitude towards fate reveals a fundamental kinship to that existential freedom which, Nietzsche insists, is man's tragedy and triumph. Chekhov sees in the individual's attitude towards "fate," whether expressed in discussion or in casual or unconscious acts, a measure of the individual's own capacity to respond to the sum total of forces acting upon him, to the **given**

---

[8] Cited by Brustein, *The Theatre of Revolt: An Approach to Modern Drama*, p. 9.
[9] F. L. Lucas, *The Drama of Chekhov, Synge, Yeats, and Pirandello* (London: Cassell, 1963), p. 41.

environment of life. If there are no on-going values but those which a man creates moment by moment while continually destroying the orientations of the past, then his freedom consists of struggling against this past in whatever form, social or psychological, the bonds assume. In *The Seagull* the characters of Nina and Treplev portray two possible responses to the given environment; the acceptance or denial of freedom.

Chekhov's attitude towards freedom and fate is most easily approached from the love angle, or perhaps we should say triangles, the dominant geometry of the play. The creative and ever-emergent personality does not passively submit itself to the love relationship; Treplev's love is distinctly not free, and he seeks to be enveloped in love, first with his mother with whom he is deadlocked in an Oedipal situation, then with Nina whom he treats as a surrogate mother. As he broods over the "failure" of his play, he complains to Nina about her supposed coldness. And later, embracing his mother after his quarrel with her, he exclaims:[10]

> Treplev: "If you only knew! I've lost everything. She doesn't love me, now I can't write. All my hopes are gone." (Act III, 44)

The connection in Treplev's mind, between creativity in art and success in love, is demonstrated to be a neurotic weakness, a mental orientation which is enslaved by its rigidity. This point is reinforced by Chekhov when he offers Nina's attitude to a similar problem:

> Nina: "...I love him, I love him ever more than before." An idea for a short story, "I love, I love passionately..." (Act IV, 68)

This is an active love; a love that is combined with readiness to seize life and hence, freedom.

The existential view here is that nothing outside the self annihilates any possibility of a ready-made future or fate; the denial of fate is an essential experience of freedom. In the play's first act, Treplev tries to determine whether his mother loves him by plucking flower-petals. Similarly in Act III, Nina tries to determine her future by fortune-telling with peas:

> Nina: "...I was trying my fortune: To be an actress or not. I wish somebody would advise me." (Act III, 38)

---

10 Anton Chekhov, "The Seagull," in *Best Plays by Chekhov*, trans. Stark Young (New York: The Modern Library, 1956), p. 44. All further references to "The Seagull" are found in this edition.

Trigorin replies as Chekhov's momentary spokesman: "In this sort of thing nobody can give advice." Chekhov realizes that there are no omens in the world; the mind must make its own connections; there is no decision-maker outside oneself. Only Nina ultimately seizes her own fate: "I have made an irrevocable decision: the die is cast; I am going on the stage, ...abandoning everything." It does not matter that her decision brings small success and great misery. What is important is that at the moment of decision she is truly free, casting out the orientations of the past and accepting a future which is but one chance after another.

Man creates meaning, he is the only real "given" in history. As Nina proclaims in the play's final glimmering of triumph:

> Nina: "...What matters is not fame, not glory, not what I used to dream about, it's how to endure, to bear my cross and have faith. I have faith and it all doesn't hurt me so much, and when I think of my calling I am not afraid of life." (Act IV, 67)

Such religious terms as "cross, faith, and calling" are particularly significant in the Nietzschean world without God. A man may choose his cross and calling; a free man has faith only in himself. That is all there is.

4. Lu Hsün's *The Biography of Ah Q*

With *The Biography of Ah Q*, we descend to a strange land of victories through the spiritual transcendence of the self. *The Biography of Ah Q* is probably best approached as an unfolding of the wills, as a continual denial to the given situations of life, or as a continuous adjustment to the Nietzschean sense of recognizing and incorporating the darker side of life as the principle of action.

Being set apart from all family and social identifications, Lu Hsün's hero exists in momentary responses to given situations. But since he is not able to cope with the sum total of forces acting upon him, his responses to various encounters are nothing but self-deceptions, or at most actions unto himself.

The emotional loophole of self-deception performs in one or both ways of self-dignifying and demeaning the opponent. It is best exemplified in the surname incident. Our hero now, being slapped and denied by Master Chao, feels despondent and angry at this turn of events. Suddenly, a sentence comes to mind: "What has the world come to? The son beats up his old man!" It sounds beautiful and powerful. He likes it, so he repeats it to himself to degrade Master Chao. Then he feels content and sings happily down the street. This is the only way he can think of re-establishing his own basic identity which has been shattered by the authority of the village. It is a spiritual self-dialectic and can be readily applied to any and all affairs, even in the realm of numbers. And it is only at this point that it can be understood why he is so elated in being grasped by the head and smashed against the wall, ultimately made to condemn himself as a beast. He is so proud of himself and so delighted with the feeling that he is the first one in the world who can condemn himself; because anything which is the "first" is great and respectable. As the meanest creature in the world and having been wholly deprived of the most basic identity as an individual, he cannot help but appeal to the paradoxical value in the concept of ordinal number to affirm his identity. The yearning for recognition as an individual and to elude reality is strong enough to allow one the conception of stratagems for protection:[11]

> "'Oh,...Here is a light bulb!' They are not afraid. No alternative.... Ah Q can only think of revenging with words. 'You are not qualified to mention it!' Then the sarcoma on his forehead at the moment becomes an honourable and dignified one to him. He is proud of it now!" (73-74)

Truly our hero from time to time would escape from the present into the past or the future; though both are uncertain to him, he would keep saying: "Ha, we were richer than you are!" or "Ha, my son will be much richer than you are!" Nevertheless, this is not a conscious illusion; it is but a stratagem for affirmation by denial, a momentary response to the reality of the present. Our hero is uprooted from the past and the future, and is doomed to respond to the given environments of every moment in the present. At this point then, one can comprehend the nature of his oblivion and endurance. The best example is found in his proposal to Widow Wu. In the kitchen, the very emblem of family life, listening to Widow Wu's gossiping about marriage affairs, he is moved; so he immediately jumps to kneel in front of the widow and utters words of desire: "I want to sleep with you! I want to sleep with you!" When the widow, shocked by such boldness, runs away screaming, he feels somehow confused; then he walks out, completely obliterating the unpleasant experience from his memory. A few minutes later, passing by the yard, he sees the widow sobbing and surrounded by neighbours and wonders what she is crying about; so out of curiosity he walks up to see what happens. It is not until he sees Master Chao rushing at him with a rod that he realizes it has something to do with him. This is the strength of his will to forget, an obscure demand that everything must be done in the right moment. And this is why every time he sees a threatening stick, he pulls himself together and waits for it, hoping that in this abuse a conclusion might be found. It is the most passive way of encountering life; an

---

[11] Lu Hsün, *The Biography of Ah Q* in *Selected Works of Lu Hsün*, Vol. I (Peking: Ren-min ch'u-pan-she, 1983), pp. 73-74. Further references to the novel are found in this edition.

orientation without action. Although when the sum total of forces acting upon him are so strong that life seems unbearable to him, he takes some action; but the object of the action is inevitably himself:

> "He raises his right hand and fiercely slaps his own face twice. It feels a little bit hot and painful, but by then he becomes calm and easy. It seems to him that the person who slaps is he himself, while the one who is being slapped is an alternative of **him**. Moments later it seems to him that he has slapped someone else, someone whom he wants to get even with. Though there is still some discomfort to his face, he lies down with contentment." (76)

The passive form of self-identification by way of encountering the given environments with a spiritual transcendence of the self climaxes with the "slapping." It is a basic aspect of human nature. It exists in all time and every place. It follows every individual as his shadow does. With Ah Q, Lu Hsün explores the darker side of life and the utter baseness of human nature. His purpose is to expose the Chinese way of life at that time and the weakness of the people.[12] But it turns out that, with his deep sympathy for the misery of life and his furious anger at the weakness of human nature, the writer has created something of value for all mankind to use in examining itself.

The tragedy of Ah Q is that he does not know what impels him to act and he has no central, constant goal for rebellion. Somehow he plays the clown to please his yearning for identity; even in parading to the execution ground, he is forced by the watching populace to sing happily in order to demonstrate his courage. He does not know what revolution is and why he is to be executed, but to console himself he thinks: "Man, born in the cosmos, is probably unavoidably to, at times, be decapitated!" These are words of truth, words of misery. Life to Ah Q is just like the mark he makes on the execution warrant;[13] it is hard for him to make it perfect or nearly round. However, in his response to life, though passive and self-deceiving, he is free and has created his own being.

5. Beckett's *Happy Days*

With Samuel Beckett and *Happy Days*, we descend to the "nothingness" of Nietzsche's wanderer, a strange no-man's land of actions contemplated but never completed. In his book, *Samuel Beckett*, Richard Coe notes that the dramatist views the human condition as "that of an indefinable **Void** within, conscious of a possible relationship with an equally indefinable **Void** without, yet invalidating

---

[12] T'ien Shun-Chuang, ed., *History of Modern Chinese Literature* (Ch'angch'un: Chilin Jen-min ch'u-pan-she, 1957), p. 214.
[13] In Chinese custom a condemned man is required to sign his own death warrant.

that relationship by the very fact of its consciousness."[14] It is that perpetual "invalidation" in which the Nietzschean "Superman" experiences the eternal joy of becoming, and in which the Beckett clown or tramp, seventy years later, finds only absurd endurance, waiting for the end.

The meaning of *Happy Days* seems mysteriously linked to the confusing notion that nothing has any real existence, and that if anything real could exist it could not be known, and that if anything were to exist and be known it could not be expressed in speech. Beckett grants existence only to the self but paradoxically uses words, the barrier of language which keeps us from knowing who and what we are, to depict the self. The dramatist agrees with Ludwig Wittgenstein (1889–1951) that one can only affirm that meaning does not exist in terms which imply that it does. Wittgenstein makes the point somewhat more concisely:

> "That the world is my world shows itself in the fact that the limits of language which I alone understand mean the limits of my world."[15]

How then can one use language after it has been devalued beyond any interpersonal relevancy? Beckett replies by paring the language to its bare non-metaphoric bones and then toning down the results through a contrapuntal relationship with action. Thus *Happy Days* is composed of equal portions of speech and stage directions. Beckett uses a staccato technique to suggest here a lack of communication: each character follows his own thoughts (in so far as Willie has any thoughts at all), while the silences and pauses isolate words and phrases, and the repetitions remind us of the tedium of this almost sub-human world.

Happy Days also presents a peculiarly modern view of time which rises from the "presentness" of existence and the denial of an historical past or definable future. Beckett offers a world near death where time has slowed almost to the zero-point only to become entangled in eternal cycles of recurrences. In a sense Winnie is buried in the "Heap of Time," the pile-up of memories and favourite objects which she uses again and again as assurances of her identity beyond the present. It is the heap which always promises yet never actually grants a death, unlike the usual experience of living in which we are continually moving to death as the framework of the present slips into the past. Winnie and Willie occupy time in the sense of a circular, finite space. It seems

---

[14] Richard Coe, *Beckett* (London: Oliver and Boyd, 1964), p. 8.
[15] Cited by Ruby Cohn, "Philosophical Fragments in the Works of Samuel Beckett" in *Samuel Beckett: A Collection of Critical Views*, ed. Martin Esslin (Englewood Cliffs: Prentice-Hall, 1965), p. 175.

that Beckett is trying to compress the linear motion of existence into a space of a few dramatic hours, in order that the present moment, in all its ordinariness, may be examined.

Through the symbiotic relationship of Willie and Winnie, the play moves toward the idea that everyone is the other's pastime; that company facilitates endurance of the pointlessness of existence, or at least conceals it by supplying a reality which becomes ever more familiar. When Winnie muses: "That is what I find so wonderful, that not a day goes by... without some addition to one's knowledge however trifling,"[16] we see Beckett undermining the rationalist's belief in a knowable reality. Knowledge in the world of *Happy Days* is impossible: it is a circle of object identification which contracts around the beholder to a fine point of absurdity, then disappears.

Happy Days is of course ironic, for the happiest day will be when one experiences the certainty of death:

> "And if for some strange reason no further pains are possible, and wait the day to come... the happiest day to come when flesh melts at so many degrees and the night of the moon has so many hundred hours.... That is what I find so comforting when I lose heart and envy the brute beast." (Act I, 18)

The brute beast is one who can die without ever considering death. As the play concludes with Willie's triumphant action towards the now helpless Winnie, we are not sure whether his movement is one of affirmation or of the final denial: the kiss or the kill.

Winnie's rational organization of her random, in this case finite, universe is in the final analysis a waste of time. It is absurd that this is the way that man **must** waste his time to maintain identity in the flux of time. It is the rational mind which is as much a cause of obscurity as the meaninglessness of the world, since his characters insist upon acting as logical beings, though logic must show them the absurdity of such behaviour. Against this background of absurdity, what then can the artist create? Beckett's answer is as close to optimism as he will allow when he offers an explanation of expression:

> "The expression that there is nothing to express, nothing from which to express, no power to express, no desire to express, together with the obligation to express."[17]

---

[16] Samuel Beckett, *Happy Days* (New York: Grove Press, 1961), p. 18. All further references to *Happy Days* are found in this edition.

[17] Cited by Arnold P. Hinchcliffe, *The Absurd: The Critical Idiom* (London: Methuen & Co., 1967), p. 67.

6. Pinter's *The Homecoming*

In his study of Harold Pinter, Walter Kerr observes that in much of the so-called "absurd" drama the playwright has reached his conclusion before beginning to write, and then has measured his illustration exactly to display that conclusion. He contends that this Platonic sequence of essence before act is a subtle betrayal of existentialist principles, a betrayal to which Pinter is not subject.[18] But we have seen that Beckett rejects process only because of its connections to plot development along a horizontal time-line, preferring instead to offer a static "present." By way of contrast, *The Homecoming* is probably best approached as an unfolding of wills; as a process or a continual becoming in the Nietzschean sense of recognizing and incorporating the darker side of life as the principle of action.

The Pinter character exists in the present and his actions are only of the present; only responses to the momentary situation. Although they may reconstruct the past from memory, they are unable to verify it or be certain of any origins except the present. For example, Max recalls his early childhood:[19]

> Max: "...He'd bend right over me, then he'd pick me up. I was only that big. Then he'd dandle me. Give me the bottle. Wipe me clean. Give me a smile. Pat me on the bum. Pass me around..." (Act I, 19)

Here, what sounds like the actual past is really vague imagining from the present. The only reality is that of the perceiver and that reality is always elusive. Pinter presents the idea as: "There are no hard distinctions between what is real and what is unreal, nor between what is true and what is false."[20] Thus the dramatist often uses the technique of casting doubt upon everything by matching each apparently clear and unequivocal statement with another contrary one having the same force.

The lack of certainty pervades the atmosphere of *The Homecoming*. Each character responds to life on private, ultimately inexplicable terms; their values and actions are not comprehensible by any external behavioral pattern or objective rational force. Thus Ruth's description of America offers only hints of bleakness which not even Teddy can understand:

---

[18] Cf. Walter Kerr, *Harold Pinter* (New York: Columbia Press, 1967), p. 8.

[19] Harold Pinter, *The Homecoming* (London: Methuen & Co. 1968), p. 19. All further references to *The Homecoming* are found in this edition.

[20] Cited by John Russel Taylor, *Anger and After: A Guide to the New British Drama* (London: Methuen & Co., 1965), p. 165.

> Ruth: "I was born quite near here. (PAUSE) Then... six years ago, I went to America. (PAUSE) It's all rock. And sand. It stretches... so far... everywhere you look. And there's lots of insects there. (PAUSE) And there's lots of insects there. (SILENCE) (Act II, 53)

The suppression of motives is inevitable because no one, not even the man who acts, can know precisely what impels him to act. According to the perpetual encountering of the Nietzschean view, man cannot be described except in terms of motion: he is what he does next; his identity is his movement.

> Ruth: "...Don't be too sure though. You've forgotten something. Look at me. I... move my leg. That's all it is. But I wear... underwear... which moves with me... it... captures your attention. Perhaps you misinterpret. The action is simple. It's a leg... moving. My lips move. Why don't you restrict... your observations to that? Perhaps the fact that they move is more significant... than the words which come through them. You must bear that... possibility... in mind." (Act II, 52-53)

Just as Teddy is rigid yet detached from objects, so Ruth is always in motion. She continues to change her identity; from model through wife, mother, daughter-in-law and sister-in-law, she evolves through the play with each encounter. No role is permitted to become absolute; even "whore" is only a possibility, never a conclusion. As a whore is what each new man wishes her to be; existentially speaking, we are all life's whores to the degree that we are in motion and have not been categorized and thereby solidified. Thus the only thing remotely certain in Pinter's world is the continuity of the encounter, for the encounter is an experience essential in confirming to the self its uniqueness and freedom. Ruth is choosing herself; she sets herself apart from all other individuals, all other beliefs and opinions, and all laws.

Lu Hsün, Beckett, and Pinter are contemporary writers; Ibsen and Chekhov belong to the previous century. Ibsenian and Chekhovian heroes are masters of their own fate. They are free to choose their own cross, for they live in a world with no absolute truth but constant efforts to find the appropriate response to every given situation. While Lu Hsün's clown secures his private identity through a passive and often self-inflicting approach to life, Beckett's tramp endures through an absurd presentness of existence, and Pinter's characters respond to life on inexplicable terms and can only be described by their motions and movements. However, they are all united in an existential viewpoint which found its most extreme expression in the work of Nietzsche. In this current age of uncertainty, it would be improper to maintain that any one of the artists discussed has answered finally the questions raised in Nietzsche's tortured confessions, for it is the nature of modern philosophy that no answer can be deemed final. The philosophical approach is especially liable to over-systematization and if we accept literature as a parameter of the "time spirit" of an age, it is hoped that future critics will begin by first recognizing a literary work of art as a personal statement, and then considering it a part of a hypothetical scheme. Furthermore, under the phenomenon of the cultural convergence, we can reaffirm a universality between Chinese and Western cyclic mentalities. As the archaic men experienced the same drive to revolt against the tyranny of time, and recognized the pattern of identification in similar celestial motion and seasonal rotation, they would inevitably adopt similar measures to guarantee personal duration against time. It is in this respect that the supposition of the cyclic myth as a universal archetype is justified.

V. CONCLUSION: CYCLIC MYTH AS INDEX OF CULTURE

A. CYCLIC ONTOLOGY VERSUS LINEAR ESCHATOLOGY

Joyce envisioned Dublin as the city which was the progress of all cities, and its circular Vico Road as the cyclic path of civilization. We will borrow his analogy and survey our own path through the complexities of a comparative analysis of Chinese and Western European literature, a path which varies in its emphasis on the cyclic or the progressive. At first this trail seemed barred by the insular homogeneity of Chinese culture, but a key was found in the relative importance and evolution of the cyclic myth in both cultures.

We have maintained that archaic man, whether Eastern or Western, shared the archetypal insight of a cyclic mentality and participated in functionally similar cyclic liturgies until the institutionalization of religion fostered divergent evolutions. The initial and essential differentiation was engendered by the influence of Hebraic Messianism in the development of Judaeo-Christian culture. By projecting the archaic victory of the king (in maintaining the periodic regeneration of all nature) into a Messianic future, Judaeo-Christianity came to perceive time and history lineally, and traced the course of humanity from initial Fall to final Redemption as a straight line. In this view, history is seen as the epiphany of God; man, created to serve and honour this divinity, achieves divine reconciliation only at the end of time. In succeeding centuries, although cyclic naturalism survived in cosmological symbolism and the cycle of the liturgical year with its solar rites of incarnation and lunar rites of atonement, death, resurrection, and ascension, Christianity remained essentially a faith of linear eschatology. This notion was assured since its very heart was the paradigmatic life of Christ, a historical event not subject to repetition.

By way of comparison, through its maintenance of the original anthropocentric emphasis and consequent lack of religious institutionalization, the heart of Chinese riteopoesis remained the ever-recurring natural cycle whose principle is the Tao which provides liberation from the anguished temporality of linear consciousness. In the Myth of Divine Administration, the cyclic myth is the divine pattern of the celestial hierarchy, with enlightenment as recognition of the Supreme Being as the ultimate divinity at the centre of a quadrantal universe of incessant regeneration. Ideally, consciousness becomes the envisagement of man's essential unity with God and his homogeneity with the divine. In the institution of Ming t'ang, the cyclic myth is the pattern of terrestrial bureaucracy, with enlightenment as recognition of the emperor as the Son of Heaven and the Representative of Earth at the centre of this tetra-symmetrical universe of perpetual renewal. Again, ideal consciousness is the envisagement of man's intermediary role in the cosmic interfusion of heaven and earth; that is to say, his position in the harmonious trinity of Heaven, Earth, and Man. At the centre of Chinese culture we find an abiding faith in this cyclic ontology.

The differentiation between linear eschatology in Judaeo-Christian culture and cyclic ontology in Chinese culture is evident and continued in the mensopoesis of the rationalization of philosophy. The history of the survival and abstraction of the cyclic mentality in Western philosophy is one of changing orientation and speculation regarding man's relationship to, rather than identity with, divinity. Thus, allowing philosophical sanction to the Hellenistic barbarian myths and mystery religions, the Neo-Platonists attempted to replace the historicity of Christ with a vulgar spiritualism which endowed events of historical Christianity with "meaning" through the importation of non-Christian natural and mythological symbolism. While the Middle Ages were dominated by the eschatological conception of the end of the world, medieval intellectuals attempted to moderate this harsh view by incorporating theories of cyclical undulation. All these attempted to render reality, through allegory and mimesis, as the operative sphere of divinity wherein the archetypes of natural cycle and mythological quest were subsumed. With the renewed anthropocentrism of the Renaissance, divine immanence was restored to the human world. Although duration beyond ephemerality still consisted of faith in God and personal redemption, and nature was seen as a temptation rather than a support, nevertheless, in the Renaissance divinity approached the immanent "suchness" of Chinese thought. In the seventeenth century a revived sense of man's limitations; his essentially linear consciousness, engendered an existence which was alienated from a corrupted nature and dependent on divinely continuous creation for duration beyond the moment. In accordance with the reductive principles of Descartes, nature and man were further estranged, consciousness became momentary sensations, and divinity achieved a mechanical "otherness" which has continued to the present. This line extends through Bacon, Descartes, Locke, Defoe, Sterne and Beckett. In the meantime, the other line extends through Rousseau, Goethe, Shelley, Yeats and Joyce: this is the attempt to restore divinity to the human world, a rehabilitation of the cyclic myth whether manifested as in the Neo-Platonic "world-soul," or in the imaginative union with the energies of nature.

At the risk of over-generalizing, the evolution and survival of the cyclic myth in Western thought presents a reliable index to the changing views of man's relationship to divinity. Thus during the medieval era the cyclic myth was subsumed within a divine hierarchy with enlightenment as recognition of God-Creator as man's only support, while during the Reformation period it was subsumed within a cosmological pattern with enlightenment as recognition of man's place in the eschatology of Christian history. Later, in the Romantic period the cyclic myth survived as the divine "spirit" of the natural world, with enlightenment as apprehension of this "spirit" immanent in the imagination and nature; while in our modern sampling, we witnessed its survival as aesthetically private patterns of history. It would appear that the predominant Christian eschatology has transformed a cyclic myth whose various survivals delineate the heterogeneity of Western European culture.

Within the Chinese mensopoesis of philosophical rationalization we do not find such heterogeneous evolution. The Chinese mind never lost sight of an essentially anthropocentric cyclic construct of a universal and perpetual oneness. Indeed, Chinese cosmology everywhere reflects the tremendous influence of the cyclic myth. Whether as the quadrantal progress of incessant renewal in the institution of the Hall of Light, or the perfect harmony of one-in-all and all-in-one in *The Book of Changes*, or the perpetual succession of five elements in sequence in the Yin-yang School, or the great current of spontaneous transformation in Taoism, or the constant process of unceasing production and reproduction in Confucianism, the Chinese universe remains a harmonious and organic whole. While the ontological reality shifts from the infinite oneness and perpetuality of an absolute divinity to the harmony of change in the ultimate Tao, then to the integral spontaneity of transformation in Nature, and finally to the wholeness of human life and the oneness of being, the Chinese epistemological belief of mutual communication and influence between the cosmos and man, between macrocosm and microcosm, remains true. The ontology becomes ever more conceptual yet existentially anthropocentric; man, with his rational mind and affectionate heart, eventually stands at the centre of the cosmos. With a sense of nostalgia, a spontaneous response, or a consciousness of responsibility, he is able to obtain freedom within a world of constant transformation and reconcile with the cosmos as a whole.

B. CYCLIC MYTH AS AN INFORMING STRUCTURE OF LITERARY WORKS AND AN INDEX OF CULTURE

Despite the above stated and implicit contrasts between the homogeneous and fragmented cultures, the literary function of the cyclic myth in Chinese and Western literature proves to be significantly similar: it is the informing structure of a literary work. In both literatures, the cyclic myth is often the underlying pattern of the temporal, spatial, and thematic structures of a particular work. The cyclic schema as a literary structure is nevertheless always pointing outside or beyond itself. Especially in Chinese literature this inclination towards transcendency seems germane to the archaic psyche which enlarged or reconciled linear consciousness by always "pointing" to the correspondences between macrocosm and microcosm. In Western literature the cyclic structure is rarely so transcendental, simply because a cyclic cosmogony has only rarely been an ontological reality for man. In any explicitly Christian frame of reference a literary work may gain universality and inclusiveness through cyclic schema, but the linear nature of consciousness and history precludes the involvement of a truly cyclic cosmogony in the work.

In its Chinese context, the cyclic schema evolves through stages of mythical envisagement, ritual manipulation and philosophical rationalization, and finds literary expression most frequently in the "quest" cycle: the call, the quest itself, and the fulfilment of the quest with reconciliation through the unity between man and the cosmos. It provides a temporal, spatial, and thematic structure consisting of the tristia in the profane time of the terrestrial world, the subsequent itineraria in the "great" time of the demonic or underworld, and the eventual reconciliation in the sacred time of the celestial sphere. As the climax of the thematic structure, reconciliation indicates the centrifugal nature of the cyclic spirit which always strives for transcendency in terms of harmony, identification, or assimilation accomplished through such literary elaborations as conversion, canonization, and immortalization. It is this transcendental nature of reconciliation, which moves beyond local literary reference to re-affirm man's unity with the cosmos, that reflects the central position of the cyclic schema in the homogeneity of Chinese mythology, ritual, philosophy, and literature.

In contrast, the Judaeo-Christian world-view of the Western literary works we have studied permits the survival of the cyclic myth in the heroic quest cycle, and in the symbolism of the natural world and the Christian liturgical year, but it is much more reluctant to posit periodicity and incessant transformation as the ontological basis of man's unity with the divine. As a master of mimesis of secular Christian reality and an encyclopedic symbolist in the exegetical tradition, Dante presents a cosmogonic hierarchy of concentric circles (a heroic quest cycle of tristia, itineraria, and reconciliation,) and an ontological hierarchy based on the poet-hero's spiralling movement to the still centre of being. What is absent in his reconciliation with divinity is the elevation of cyclic Nature, as the immediate field of divine immanence, to a position of mutual interdependence between Man and God. *The Commedia* is cyclic in structure, but linear and progressive insofar as reconciliation lies in the future beyond the corrupted mundane world. Focusing on the tristia, the linear and instantaneous consciousness of man, Milton's *Paradise Lost* affirms man's position in the corrupted world as a consequence of the parallel fall of Man and Satan. Here, continuous creation in the present and divine reconciliation in the future become the justification for the linearity of time and history; the natural world is seen as a reflection of man's own limitations. Although maintaining the progressive view of history, Goethe challenges the limitations of linear consciousness by rehabilitating cyclic nature to a position of immanence, his "world-soul." Yet so strong is the Cartesian awareness of sensitive momentary consciousness that Faust must deny the possibility of permanent value within natural flux even as the "valued" moment appears. Romantic thought of the late eighteenth and early nineteenth centuries continued Goethe's dissection of the moment, and most closely approached the Chinese world-view in the rehabilitation of cyclic nature as divine immanence accessible to and paradigmatic in the human imagination. But ever since the rise of a scientific materialism which tended to regard nature as an object for analysis and exploitation rather than a field of participation, the cyclic harmony of Man, God, and Nature has never achieved any ontological significance in the popular Western mind; despite the warning of an inevitable alienation and absurd endurance in the linear consciousness in the works of Beckett, Cohen and Pinter, and the renewal of cyclic concepts of history in the works of Yeats and Joyce.

Considering the increasing popularity of the cyclic schema in recent years, and its survival in China for upwards of a millennium, it may yet prove to be the most satisfactory ontological position for the individual psyche against the linear consciousness of a world of change. Even as the central tenet of the homogeneous Oriental culture it has not remained static, and in Chinese thought we have traced the shifting ontological reality from the divine to the natural and eventually to the existential. Certainly in the West the retreat of the cyclic myth seemed inevitable with the abandonment of the archetypes of repetition, while in the anthropocentric culture of the Chinese world its dispersion was held in check by the counterpoise of the terrestrial and the celestial, and the mediating role of man in the interfusion of Heaven and Earth. It is sustained through the concept of **"T'ien Jen Ho I"**, the unity of Heaven, Earth and Man.

A student approaching the comparative study of Chinese literature must acknowledge the unique homogeneity of man, nature, and divinity in terms which have only recently been approached by Western literature. He must acknowledge the continuing conviction that there is a mutual communication or influence between the cosmos and man who stands at the centre of universal events. And he must acknowledge that man is then responsible for maintaining the cosmic harmony, a fact which delineates the notion that the Chinese view not only rejects the destiny of human beings as final and irreducible, but also rejoices in the transcendental freedom within a world of perpetual change and universal wholeness. It is this, man's creative ability on the cosmic plane as well as the historical, which most distinguishes Chinese man from his Western counterpart.

BIBLIOGRAPHY

Albright, Daniel. *The Myth Against Myth: A Study of Yeats' Imagination in Old Age.* London: Oxford University Press, 1972.
Anonby, August J. *Milton's View of Human Destiny.* Vancouver: University of British Columbia Press, 1965.
Auerbach, Erich. *Dante: Poet of the Secular World.* Tr. Ralph Mannheim. Chicago & London: The University of Chicago Press, 1961.
Bates, Paul A. *Faust: Sources, Works, Criticism.* New York: Harcourt, Brace & World, 1969.
Bates, Ronald. *Northrop Frye.* Toronto: McClelland and Stewart, 1971.
Beckett, Samuel. *Happy Days.* New York, Grove Press, 1961.
_____. *Watt.* New York: Grove Press Inc., 1959.
Bergin, Thomas Goddard. *Dante's Divine Comedy.* Englewood Cliffs: Prentice-Hall, 1971.
_____. Ed. *From Time to Eternity.* New Haven & London: Yale University Press, 1967.
Benstock, Bernard. *Joyce-Again's Wake: An Analysis of Finnegans Wake.* Settle & London: University of Washington Press, 1965.
Binyon, Laurence. *The Spirit of Man in Asian Art.* New York: Dover Publication, 1965.
Birch, Cyril. *Anthology of Chinese Literature.* New York: Grove Press, 1972.
_____. *Studies in Chinese Literary Genres.* Berkeley & London: University of California Press, 1974.
Bloom, Harold. Ed. *Romanticism and Consciousness: Essays in Criticism.* New York: W.W. Norton & Company, 1970.
_____. *The Visionary Company: A Reading of English Romantic Poetry.* London: Faber & Faber, 1961.
Bochenski, I. M. *Contemporary European Philosophy.* Berkeley: University of California Press, 1956.
Bodde, Derk. *Essays on Chinese Civilization.* Ed. Charles Le Blanc and Dorothy Borei. Princeton: Princeton University Press, 1981.
_____. "Evidence for 'Laws of Nature' in Chinese Thought," in *Harvard Journal of Asiatic Studies.* 20-3, 4, Dec., 1957.
Bodkin, Maud. *Archetypal Patterns in Poetry: Psychological Studies of Imagination.* London: Oxford University Press, 1934.
Boone, William J. And Legge, James. *The Notions of the Chinese Concerning God and Spirits.* Taipei: Cheng Wen Publishing Co., 1971.
Bouchard, Donald F. *Milton: A Structural Reading.* Montreal: McGill-Queen's University Press, 1974.
Brisman, Leslie. *Milton's Poetry of Choice and Its Romantic Heirs.* Ithaca: Cornell University Press, 1973.
Browne, Robert M. "The Typology of Literary Signs," *College English,* XXXIII, (1971), pp.1-17.
Brustein, Robert. *The Theatre of Revolt: An Approach to Modern Drama.* New York: Atlantic Press Monthly Book, 1969.
Burkman, Katherine H. *The Dramatic World of Harold Pinter.* Ohio State University Press, 1971.
Bush, Susan. "The Chinese Literati on Painting: Su Shih to Tung Ch'i-Ch'ang," *Harvard-Yenching Institute Studies,* XXVII. Cambridge: Harvard University Press, 1971.
Butor, Michel. *Inventory.* Ed. Richard Howard. New York: Simon and Schuster, 1968.

Campbell, Joseph. *The Hero with a Thousand Faces.* New York: Pantheon Books, 1949.
_____. *The Masks of God: Oriental Mythology.* New York: The Viking Press, 1972.
Camus, Albert. *The Myth of Sisyphus.* Tr. Justin O'Brien. London: H. Hamilton, 1955.

Cassirer, Ernst. *Language and Myth*. Tr. Susanne K. Langer. New York: Dover Publications, 1953.
Chan An-tai. *Ch'ü Yüan*. Shanghai: Jen-min ch'u-pan-she, 1957.
Chan Wing-tsit. *A Source Book in Chinese Philosophy*. Princeton: Princeton University Press, 1972.
_____. Ed. *The Great Asian Religion: An Anthology*. London: The Macmillan Co., 1969.
Chandler, S. Bernard. Ed. *The World of Dante: Six Studies in Language and Thought*. Toronto: University of Toronto Press, 1966.
Chang Chih. *T'ao Yüan-ming chuan lun*. Shanghai: Hsüeh-ti ch'u-pan-she, 1953.
Chang H.C. *Chinese Literature*. Edinburgh: Edinburgh University Press, 1973.
Chang Han-Liang. "The Archetypal Structure of the Circle of 'Yang-lin Story'," in *Chung-kuo ku-tien wen-hsüeh lun-ts'ung III: On Myth and Fiction*. Taipei: Chung-wai wen-hsüeh ch'u-pan-she, 1976.
Chang Kuang-chih. "A Classification of Shang and Chou Myths," in *Bulletin of the Institute of Ethnology, Academia Sinica*, 14. Taipei, 1962.
_____. "The Chinese Creation Myths: A Study in Method," in *Bulletin of the Institute of Ethnology, Academia Sinica*, 8. Taipei, 1959.
Chao Ts'ung. *Chung-kuo ssu ta hsiao-shuo chih yen-chiu*. Hong Kong: You-lien ch'u-pan-she, 1964.
Chase, Richard. *Quest for Myth*. Baton Rouge: Louisiana State University Press, 1949.
Chekhov, Anton. *The Seagull*. Tr. Stark Young. New York: The Modern Library, 1956.
Chen Charles K. H. Ed. *Neo-Confucianism: Essays by Wing-tsit Chan*. New York: Oriental Society, 1969.
Chen Chung-yü. "An Analysis of the Dragon Design on the Bone-Artifacts of the Shang Dynasty," in *Bulletin of the Institute of History and Philology, Academia Sinica*, 41-3. Taipei, 1969.
Chen Ting. *Anthology of Inscriptions on Oracle Bones*. Taipei: Ta-t'ung ch'u-pan-she, 1971.
Cheng Heng-hsiung. "Metamorphosis in Mythology: A Comparative Study of Greek and Bunun Mythology," *Tamkang Review*, II-2, III-1. Taipei: Tamkang College Press, 1971-72.
Ch'ien Mu. *Chung-kuo wen-hua-shih tao-lun*. Taipei: Cheng-chung shu-chü, 1972.
_____. "The Conception of Spirits and Deities in the History of Chinese Thought," in *The New Asia Journal*, I-1. Hong Kong: NAJ, 1955.
Chou He. *Ch'un-ch'iu chi-li k'ao-pien*. Taipei: Chia-hsin wen-hua chi-chin-hui, 1970.
Christiani, Dounia. *The Wild Duck: A New Translation, The Writing of the Play*, Criticism. New York: W. W. Norton, 1968.
Chu Dan, *The Creative Art of Lu Hsün*. Shanghai: Hsin-yi ch'u-pan-she, 1958.
Ch'ü Wan-li. "On the Date of the kua-tz'u and yao tz'u of I Ching," in the *Bulletin of the College of Arts*, 1. Taipei: National Taiwan University Press, 1950.
_____. "Plastronmancy as the Original Source of I Ching," in *Bulletin of the Institute of History and Philology, Academia Sinica*, XXXIII. Taipei, 1962.
Clark, Robert Thomas Rundle. *Myth and Symbol in Ancient Egypt*. London: Thames & Hudson, 1959.
Coe, Richard N. *Beckett*. London: Oliver and Boyd, 1964.
_____. *Samuel Beckett*. New York: Grove Press Inc., 1968.
Cohen, Leonard. *Beautiful Losers*. New York: The Viking Press, 1967.
Cohn, Ruby. *Currents in Contemporary Drama*. Bloomington: Indiana University Press, 1969.
_____. *Samuel Beckett: The Comic Gamut*. New Brunswick & New Jersey: Rutgers University Press, 1962.
Cullmann, Oscar. *Christ and Time: The Primitive Christian Conception of Time and History*. Tr. Floyd V. Filson. London: SCM Press, 1952.

Defoe, Daniel. *Moll Flanders*. Ed. James Sutherland. Boston: Houghton Mifflin Company, 1959.
Djwa, Sandra. "Leonard Cohen, Black Romantic," *Canadian Literature*, 34 (1967).
Demaray, John G. *The Invention of Dante's Commedia*. New Haven, & London: Yale University Press, 1974.
Dillistone, Frederick William. *Myth and Symbol*. London: S.P.C.K., 1968.
Doane, T. W. *Bible Myths and Their Parallels in Other Religions*. New York: University Books, 1971.
Dobbins, Austin C. *Milton and the Book of Revelation: The Heavenly Cycle*. University: University of Alabama Press, 1975.
Dudbridge, Glen. *The Hsi Yu Chi: A Study of Antecedents to the Sixteenth Century Chinese Novel*. Cambridge (Eng.) University Press, 1970.
Dunbar, H. Flanders. *Symbolism in Medieval Thought and Its Consummation in the Divine Comedy*. New York: Russel & Russel, 1961.
Duncan, J. Ellis. *Milton's Earthly Paradise: A Historical Study of Eden*. Minneapolis: University of Minnesota Press, 1972.
Eastman, Arthur M. Ed. *The Norton Anthology of Poetry*. New York: W. W. Norton & Commpany Inc. 1970.
Eberhard, Walfram. Ed. *The Man in the Moon: Folktales of China*. Chicago: The University of Chicago Press, 1968.
Eliade, Mircea. *Cosmos and History: The Myth of the Eternal Return*. New York: Harper Torch Books, 1959.
_____. *Myth, Dream and Mysteries*. Tr. Philip Mairet. New York & Evanston: Harper & Row Publisher, 1967.
_____. *Myth and Reality*. New York & Evanston: Harper & Row Publishers, 1963.
Eliot, T.S. *Dante*. London: Faber & Faber, 1929.
Esslin, Martin. *Samuel Beckett: A Collection of Critical Essays*. Englewood Cliffs: Prentice-Hall, 1965.
_____. *The Theatre of the Absurd*. London: Eyre & Spottiswoodie, 1966.
Fang Shan-chu. "K'un-lun t'ien-shan yü t'ai-yang-shen," The Continent Magazine, 49-4. Taipei, 1974.
Fang Thomé H. "The World and the Individual in Chinese Metaphysics," in *The Chinese Mind*. Ed. Charles A. Moore. Honolulu: University of Hawaii Press, 1967.
Fann. K.T. Ed. *Ludwig Wittgenstein: The Man and His Philosophy*. New York: Dell Publishing Co., 1967.
Feder, Lillian. *Ancient Myth in Modern Poetry*. Princeton: Princeton University Press, 1971.
Fergusson, Francis. *Dante*. New York & London: The McMillan Company, 1966.
_____. *Dante's Drama of the Mind: A Modern Reading of the Purgatario*. Princeton: Princeton University Press, 1953.
Ferguson, John C. "Chinese Mythology," in *The Mythology of All Races*, III. Ed. Canon MacCulloch. New York: Cooper Square Publishers, 1964.
Fjelde, Rolf. Ed. *Ibsen: A Collection of Critical Essays*. Englewood Cliffs: Prentice-Hall, 1965.
Fletcher, Angus. *Allegory: Theory of Symbolic Mode*. Ithaca: Cornell University Press, 1964.
Fletcher, Jefferson B. *Symbolism of the Divine Comedy*. New York: Columbia University Press, 1921.
Fordham, Frieda. *An Introduction to Jung's Psychology*. Maryland: Penguin Books, 1953.
Frazer, James G. *The Golden Bough: A Study in Magic and Religion*. London: Macmillan, 1920.
Frye, Northrop. *Anatomy of Criticism*. Princeton: Princeton University Press, 1957.
_____. *A Study of English Romanticism*. New York: Random House, 1968.
_____. *Fables of Identity: Studies in Poetic Mythology*. New York: Harcourt, Brace & World, 1963.

_____. *Fools of Time: Studies in Shakespearean Tragedy*. Toronto: University of Toronto Press, 1967.
_____. *The Return of Eden: Five Essays on Milton's Epics*. Toronto: University of Toronto Press, 1965.
Fudan University. *History of Modern Chinese Literature*. Fudan University Press, 1969.
Fung Yu-lan. *A Short History of Chinese Philosophy*. Ed. Derk Bodde. New York: The Free Press, 1966.
Fu Yün-sen. *A Study on the Decimal System of Dating*. Hong Kong: Chung-shan ch'u-pan-she, 1972.
Gardner, Martin. "Mathematical Games," *Scientific American*, 230-1. New York, 1974.
Gaster, Theodor Herz I. *Myth, Legend, and Custom in the Old Testament: A Comparative Study with Chapters from Sir James G. Frazer's Folklore in the Old Testament*. New York: Harper & Row, 1969.
Ghent, Dorothy Van. *The English Novel*. New York: Harper & Row Publishers, 1967.
Gilbert, Allan H. *Dante and His Comedy*. New York: New York University Press, 1963.
Gilson, Etienne Henry. *Dante and Philosophy*. Tr. David More. London: Sheed & Ward, 1952.
Glasser, Richard. *Time: In French Life And Thought*. Tr. C. G. Pearson. New Jersey: Manchester University Press, 1972.
Graham, A.C. Tr. *The Book of Lieh Tzu*. London: John Murray, 1960.
Gross, John. *Joyce*. Fontana: Collins Sons & Co., 1971.
Grose, Christopher. *Milton's Epic Process: Paradise Lost and Its Miltonic Background*. New Haven: Yale University Press, 1973.
Gunn, Edward M. Ed. *Twentieth-Century Chinese Drama: An Anthology*. Bloomington: Indiana University Press, 1983.
Halbrooks, Richard Thayer. *Dante and the Animal Kingdom*. New York: AMS Press, 1966.
Hassan, Ihan H. "Toward a Method in Myth," in *Journal of American Folklore*, 65, 1952.
Hawkes, David. Tr. Cao Xueqin. *The Story of the Stone*. Penguin Books, 1973.
_____. *Ch'u Tz'u: The Songs of the South*. Boston: Beacon Press, 1962.
_____. "The Quest of the Goddess," in *Asia Major*, XIII-1, 2, 1967.
Hesla, David H. *The Shape of Chaos: An Interpretation of the Art of Samuel Beckett*. Minneapolis: University of Minnesota Press, 1971.
Hightower, James Robert. *The Poetry of T'ao Ch'ien*. Oxford: Clarendon Press, 1970.
Hinchcliffe, Arnold P. *The Absurd: The Critical Idiom*. London: Methuen & Co., 1967.
_____. *Harold Pinter*. New York: Twayne Publishers Inc., 1967.
Hinks, Roger Packman. *Myth and Allegory in Ancient Art*. Nendela, Liechtenstein, Kraus reprint, 1968.
Hoeffer, Jacqueline. "Watt," in *Samuel Beckett: A Collection of Critical Essays*. Ed. Martin Esslin. Englewood Cliffs: Prentice-Hall Inc., 1965.
Hoffman, Daniel G. *Barbarous Knowledge: Myth in the Poetry of Yeats, Graves and Muir*. New York: Oxford University Press, 1967.
Hoffman, Frederick J. *Samuel Beckett: The Language of Self*. New York: E. P. Dutton Co., 1964.
Ho Kuan-Chou. "The Shan-hai ching: The Date of Its Author and its Scientific Value," in *Yenching Journal of Chinese Studies*, 7. Peking: Yenching University Press, 1930.
Hollander, Robert. *Allegory in Dante's Commedia*. Princeton: Princeton University Press, 1969.
Holtan, Orley I. *Mythic Patterns in Ibsen's Last Plays*. Minneapolis: University of Minnesota Press, 1970.
Hou Wai-lu. *Lun T'ang Hsien-tsu chü-tsuo ssu chung*. Peking: Chung-kuo hsi-chü ch'u-pan-she, 1062.
Hsia C. T. and Hsia T. A. "New Perspective on Two Ming Novels: Hsi Yu Chi and Hsi Yu Pu," in *Wen Lin: Studies in the Chinese Humanities*. Ed. Chow Tse-tsung. Milwaukee and London: University of Wisconsin Press, 1968.

Hsia C. T. *The Classic Chinese Novel*. New York & London: Columbia University Press, 1968.
_____. "Time and Human Condition in the Plays of T'ang Hsien-tsu," in *Self and Society in Ming Thought*. Ed. Theodore de Bary. New York & London: Columbia University Press, 1970.
Hsiao Wang-ch'ing. *T'ao Yüan-ming p'i-p'ing*. Taipei: Kai-ming shu-chü, 1974.
Hsü Shih-nien. *Essays on Classic Chinese Novel*. Shanghai: Shanghai shu-chü, 1955.
Huang Wen-shan. "Totemism and the Origin of Chinese Philosophy," in *Bulletin of the Institute of Ethnology, Academia Sinica*, 9. Taipei, 1960.
Hung Josephine Huang. *Ming Drama*. Taipei: Heritage Publishing Co., 1966.
Hunt, Theodore W. *Literature: Its Principles and Problems*. New York: Funk & Wagnalls Co., 1906.
Hu Shih. "The concept of Immorality in Chinese Thought," in *Bulletin of the Institute of History and Philology, Academia Sinica*, XXXIV-2, Taipei, 1963.
Ibsen, Henrik. "The Wild Duck" in *Henrik Ibsen: Four Major Plays*, Tr. and Ed. Rolf Fjelde, New York: New American Library, 1965.
Jackson, Robert Louis, Ed. *Chekhov: A Collection of Critical Essays*. Englewood Cliffs: Prentice-Hall, 1967.
Jameson, R. D. *Three Lectures on Chinese Folklore*. Peking: San Yu Press, 1932.
James, Edwin Oliver. *Myth and Ritual in the Ancient Near East: An Archaeological and Documentary Study*. London: Thames & Hudson, 1958.
Jaspers, Karl. *Myth and Christianity: An Inquiry into the Possibility of Religion without Myth*. New York: Noonday Press, 1958.
Jeffares, A. Norman. *W. B. Yeats: Selected Poetry*. London: The MacMillan Company, 1967.
Jen Yu-wen. "Ch'en Hsien-chang's philosophy of the Natural," in *Self and Society in Ming Thought*. Ed. Theodore de Bary. New York: Columbia University Press, 1970.
Jonas, Hans. *The Gnostic Religion: The Message of the Alien God and the Beginning of Christianity*. Boston: Beacon Press, 1963.
Joyce, James. *Finnegans Wake*. New York: The Viking Press, 1939.
Jung, C. G. *Modern Man in Search of a Soul*. Tr. W. S. Dell and Cary F. Baynes. New York: Harcourt, Brace & World, 1962.
Kang P'ing. "Lun Wei-Chin yu-hsien-shih te hsing-suai yü lei-pieh," in *Chung-wai wen-hsüeh*, 3-5. Taipei: Chung-wai wen-hsüeh ch'u-pan-she, 1974.
Kastor, Frank S. *Milton and the Literary Satan*. Amsterdam: Rodopi, 1974.
Kenner, Hugh. *Samuel Beckett: A Critical Study*. Berkeley and Los Angeles: University of California Press, 1968.
Kermode, Frank. *Puzzle and Epiphanies*. London: Routledge & Kegan Paul, 1963.
_____. *The Sense of an Ending*. New York: Oxford University Press, 1967.
Kerr, Walter. *Harold Pinter*. New York: Columbia University Press, 1967.
Kirk, Geoffrey Stephen. *Myth: Its Meaning and Functions in Ancient and Other Cultures*. Berkeley: University of California Press, 1970.
Kitagawa, Joseph M. and Long, Charles H. Ed. *Myths and Symbols: Studies in Honour of Mircea Eliade*. Chicago & London: The University of Chicago Press, 1969.
Knight, George Wilson. *Myth and Miracle: An Essay on the Mythic Symbolism of Shakespeare*. London: Burrow & Co., 1929.
Knott, J. Ray. *Milton's Pastoral Vision: An Approach to Paradise Lost*. Chicago: University Of Chicago Press, 1971.
Knox, John. *Myth and Truth: An Essay on the Language of Faith*. Charlottesville: University Press of Virginia, 1964.
Koestler, Arthur. *The Act of Creation*. London: Pan Books Ltd., 1969.
Krieger, Murray. *Northrop Frye in Modern Criticism*. New York & London: Columbia University Press, 1966.

Kuan Tung-kuei. "On Chinese Ancient Myths of the Rotational and the Simultaneous Appearances of the Ten Suns," in *Bulletin of the Institute of History and Philology, Academia Sinica*, XXXIII. Taipei, 1962.

_____. "The Harvest Festival of Ancient China and Its Relation to the Calendrical Year," in *Bulletin of History and Philology, Academia Sinica*, XXXI. Taipei, 1960.

Ku Chieh-kang. *Wu-te chung-shih shuo hsia te cheng-chih He li-shih*. Hong Kong: Lung-men shu-chü, 1970.

Kuo Mo-jo. *Ch'ü Yüan yen-chiu*. Ch'ün-yi ch'u-pan-she, 1946.

_____. *The Epoch of the Elaboration of the Book of Changes*. Ch'angsha: Shang-wu yin-shu-kuan, 1940.

Kuo Ting-T'ang. *The Development of the Concept of Heaven in Pre-Ch'in Period*. Shanghai: Shang-wu yin-shu-kuan, 1936.

Kurth, Burton O. *Milton and Christian Heroism*. Berkeley: University of California Press, 1966.

Lahr, John. Ed. *A Casebook on Harold Pinter's The Homecoming*. New york: Grove Press Inc., 1971.

Lang, Andrew. *Myth, Ritual, and Religion*. New York: AMS Press, 1968.

Langer, Susanne K. *Feeling and Form*. London: Routledge Kegan Paul Ltd., 1967.

_____. *Philosophy in A New Key*. New York & London: A Mentor Book, 1951.

Lao Kan. "The Division of Time in the Han Dynasty as Seen in the Wooden Slips," in *Bulletin of the Institute of History and Philology, Academia Sinica*, XXXIX-1. Taipei, 1969.

Lee Chia-yen. "Ch'ü Yüan li-sao ssu-hsiang he yi-shu," in *Ch'u-tz'u yen-chiu lun-wen chi*, III. Chung-kuo yi-wen ch'u-pan-she, 1969.

Lee Han-san. *Hsien-Ch'in liang-Han chih yin-yang wu-hsing hsüeh-shuo*. Taipei: Chung-ting wen-hua ch'u-pan-she, 1967.

Legge, James. *Religions of China*. London: Hadder and Stoughton, 1880.

_____. *The Sacred Book Of China: The I Ching*. New York: Dover Publication, 1964.

Levin, Harry. *James Joyce: A Critical Introduction*. New York: New Directions Publishing Co., 1941.

Li Chi. *Beginning of Chinese Civilization*. Seattle: University of Washington Press, 1957.

Li Tsung-t'ung. "New Interpretations of Yen Ti and Huang Ti," in *Bulletin of the Institute of History and Philology, Academia Sinica*, XXXIX-1. Taipei, 1969.

Lin Hen-li. "A Behaviouristic Study of the Creation Myth," in *Bulletin of the Institute of Ethnology, Academia Sinica*, 14. Taipei, 1962.

Lin Hui-hsiang. *Shen-hua lun*. Taipei: Shang-wu yin-shu-kuan, 1968.

Lin Wen-keng. *History of Chinese Literature*. Taipei: Kuang-wen shu-chü, 1963.

Lin Wen-yüeh. "Ts'ung yu-hsien-shih tao shan-shui-shih," in *Chung-wai wen-hsüeh*, 1-9. Taipei, 1973.

Liu Chieh. *Chung-kuo ku-tai tsung-tsu i-chih shih lun*. Taipei: Cheng-chung shu-chü, 1971.

Liu Hsiu-yeh. *Wu Ch'eng-en shih-wen chi*. Shanghai: Ku-tien wen-hsüeh ch'u-pan-she, 1958.

Liu Wu-chi. *An Introduction to Chinese Literature*. Bloomington & London: Indiana University Press, 1966.

Liu Yüan-lin. "On the Relationship of Fu Hsi and Nü Wa in the Mythology to the 'K'un' in the Oracle Bone Inscription," in *Bulletin of the Institute of history and Philology, Academia Sinica*, XLI-4. Taipei, 1969.

Locke, John. *An Essay Concerning Human Understanding* (Great Books of Western World: Volume 35, Bk. ii, Ch.33-5). Chicago: Encyclopedia Britannica, Inc, 1952.

Lucas, F. L. *The Drama of Chekhov, Synge, Yeats, and Pirandello*. London: Cassell, 1963.

Luh Ch'in-li. "T'ao Yüan-ming's Poem 'Form, Shadow and Soul' and the Doctrine of Buddhism and Taoism in the East Chin," in *Bulletin of the Institute of History and Philology, Academia Sinica*, VII-3. Shanghai: Shang-wu yin-shu-kuan, 1937.

Lu Hsün. *A Brief History of Chinese Fiction.* Tr. Yang Hsien-yi & Gladys Yang. Peking: Foreign Language Press, 1958.

_____. "The Biography of Ah Q," in *Selected Works of Lu Hsün.* Peking: Jen-min ch'u-pan-she, 1983.

Lyons, Charles R. *Henrik Ibsen: The Divided Consciousness.* Carbondale: Southern Illinois University Press, 1972.

Malinowski, Bronislaw. *Myth in Primitive Psychology.* Westport, Conn.: Negra University Press, 1971.

Marilla, E. Lingworth. *Milton and Modern Man.* University: University of Alabama Press, 1968.

Mao Tun. *Shen-hua tza-lun.* Shanghai: Shih-chie shu-chü, 1929.

Max Maltenmark. *Lao Tzu and Taoism.* Tr. Roger Greaves. Stanford: Stanford University, 1969.

Mchugh, Florence and Isabel. Tr. *The Dream of the Red Chamber.* New York: Pantheon Books, 1958.

McLuhan, Marshall. *Understanding Media: The Extensions of Man.* New York: McGraw-Hill Book Co., Paperback Edition, 1965.

Mendilow, A. A. *Time And The Novel.* New York: Humanities Press, 1965.

Meng Yao. *History of Chinese Fiction.* Taipei: Biographical Literature Press, 1969.

Merivale, Patricia. *Pan The Goat-God: His Myth in Modern Times.* Cambridge: Harvard University Press, 1969.

Meyerhoff, Hans. *Time In Literature.* Berkeley and Los Angeles: University of California Press, 1968.

Middleton, John. *Myth and Cosmos: Reading in Mythology and Symbolism.* Garden City, N.Y.: The National History Press, 1967.

Miller, James Edwin. *Myth and Method: Modern Theories of Fiction.* Lincoln: University of Nebraska Press, 1960.

Miller, Lucien. "Poetry as Contemplation: T'ao Ch'ien's Homing and William Wordsworth's Tintern Abby," in *Journal of the Institute of Chinese Studies of the University of Hong Kong,* 6-2, Hong Kong, 1973.

Moore, Charles A. *The Chinese Mind.* Honolulu: University of Hawaii Press, 1967.

Murillo, Louis Andrew. *The Cyclical Night: Irony in James Joyce and Jorge Luis Borges.* Cambridge: Harvard University Press, 1968.

Murray, Henry Alexander. *Myth and Mythmaking.* New York: G. Braziller, 1960.

Murray, Patrick. *Milton: The Modern Phase.* London: Author, 1967.

Needham, Joseph. *Time and Eastern Man.* The Henry Myers Lecture 1964. Royal Anthropological Institute: Occasional Paper No. 21, 1965.

_____. *Time: The Refreshing River.* London: George Allen & Unwin Ltd., 1948.

Okada, Takehiko. "Wang Chi and the Rise of Existentialism," in *Self and Society in Ming Thought.* New York & London: Columbia University Press, 1970.

Ondaatje, Michael. *Leonard Cohen.* Toronto: McClelland & Stewart, 1970.

Ou Hsieh-fang. "T'ang Hsien-tsu chi ch'i huan-hun chi," in *The Continent Magazine,* XXII-9, 10, 11. Taipei, 1961.

Pacey, Desmond. "The Phenomenon of Leonard Cohen," *Canadian Literature,* 34, (1967).

Patai, Raphael. *Myth and Modern Man.* Englewood Cliffs: Prentice-Hall, 1972.

Peckham, Morse. *Beyond the Tragic Vision.* New York: George Brazilier, 1962.

_____. *Man's Rage for Chaos: Biology, Behaviour, and the Arts.* New York: Schoken Books, 1967.

Pi Hai-chi. *Wen-hsüeh yen-chiu hsü-chi.* Taipei: Shang-wu yin-shu-kuan, 1971.

_____. *T'ang Hsien-tsu ch'uan-ch'i chih yen-chiu.* Taipei: Shang-wu yin-shu-kuan, 1974.

Pinter, Harold. *The Homecoming.* London: Methuen & Co., 1965.

Poulet, Georges. *Studies in Human Time.* Tr. Eliot Coleman. Baltimore: Johns Hopkins Press, 1956.

_____. "The Metamorphosis of the Circle," in *Dante: A Collection of Critical Essays*. Ed. John Freccero. Englewood Cliffs: Prentice-Hall, 1965.
Priestley, J. B. *Man and Time*. London: Aldus Books, 1968.
Rank, Otto. *The Myth of the Birth of the Hero and Other Writings*. Ed. Philip Freund. New York: Vintage Books, 1959.
Reik, Theodor. *Myth and Guilt: The Crime and Punishment of Mankind*. New York: G. Braziller, 1957.
Richardson, Samuel. *Pamela*. London: J. M. Dent & Sons Ltd., 1959.
Riencourt, Amaury de. *The Soul of China*. New York: Coward-McCann Inc., 1958.
Romilly, Jacqueline de. *Time in Greek Tragedy*. Ithaca: Cornell University Press, 1968.
Ross, Stephen D. *Literature and Philosophy*. New York: Appleton-Century-Crofts, 1969.
Rousselle, Erwin. "Dragon and Mare: Figures of Primordial Chinese Mythology," in *The Mystic Vision*. (Paper from the Eranos Yearbooks, 6.) London: Routledge & Kegan Paul, 1968.
Rowley, George. *Principles of Chinese Painting*. Princeton: Princeton University Press, 1959.
Samuel, Irene. *Dante and Milton: The Commedia and Paradise Lost*. Ithaca: Cornell University Press, 1966.
Savage, Giles Christophor. *Cycle of Time and Seasons Based on the Reproduced Ancient Hebrew Calendar*. Nashville, Tnn.: Sunday School Board of the Southern Baptist Convention, 1928.
Schafer, Edward H. *The Divine Woman: Dragon Ladies and Rain Maidens in T'ang Literature*. San Francisco: North Point Press, 1980.
Schefold, Karl. *Myth and Legend in Early Greek Art*. Tr. Audrey Hicks. New York: H.N. Abrams, 1966.
Sebeok, Thomas Albert. Ed. *Myth: A Symposium*. Bloomington: Indiana University Press, 1965.
Shan-hai ching, t'u-tzan, pu-chu. Taipei: Hsin-lu shu-chü, 1927.
Shapiro, Marianne. *Woman Earthly and Divine in the Comedy of Dante*. Lexington: The University of Kentucky Press, 1975.
Shih Sheng-yen. *Comparative Religion*. Taipei: Chung-hua shu-chü, 1970.
Shumsker, Wayne. *Literature and the Irrational: A Study in Anthropological Background*. Englewood Cliffs: Prentice-Hall, 1960.
Sie Kang. "A Preliminary Study of Chinese Magic from the Sociological Standpoint," in *Tunghai Journal*, VIII-2, Taichung, 1967.
Simonson, Harold P. *Strategies in Criticism*. New York: Holt, Rinehart and Winston, 1971.
Sisson, Charles Jasper. *The Mythical Sorrows of Shakespeare*. London: H. Milford, 1934.
Sivin, Nathan. "Chinese Conception of Time," in *The Earlham Review*, I, Fall, 1966.
Slochower, Harry. *Mythopoesis: Mythic Patterns in the Literary Classics*. Detroit: Wayne State University Press, 1970.
Slote, Bernice. Ed. *Myth and Symbol: Critical Approaches and Applications by Northrop Frye, L.C. Knights and Others*. Lincoln: University of Nebraska Press, 1963.
Soothill, William Edward. *The Hall of Light*. London: Lutterworth Press, 1951.
Spenser, Sharon. *Space, Time and Structure in the Modern Novel*. New York: Grove Press Inc., 1965.
Stallknecht, Newton P. and Horst Frenz. Ed. *Comparative Literature: Method and Perspective*. Carbondale: Southern Illnois University Press, 1961.
Steiner, George. *Language and Silence*. New York: Atheneum, 1967.
Strelka, Joseph. *Perspective in Literary Symbolism*. University Park: Pennsylvania State University Press, 1968.
Sterne, Laurence. *Tristram Shandy*. Ed. James A. Work. New York: The Odyssey Press, 1940.
Su Hsüeh-lin. *Chiu-ko chung jen-shen lien-ai wen-ti*. Taipei: Wen-hsing ch'u-pan-she, 1967.

Tai Chen. "Ch'ü Yüan fu chu," in *Ch'u-tz'u ssu chung*. Hong Kong: Kuang-chih ch'u-pan-she, 1959.
Tai Chün-jen. "The Essence and Original Function of the Ho-t'u luo-shu," in *Bulletin of the College of Arts*, 15. Taipei: National Taiwan University Press, 1966.
T'an Cheng-pi. *Chung-kuo wen-hsüeh-chia ta tz'u-tien*. Hong Kong: Wen-shih ch'u-pan-she, 1934.
T'ang Chün-i. *Chung-kuo wen-hua chih ching-shen chia-chih*. Taipei: Cheng-chung shu-chü, 1972.
_____ . "The Conception of the Will of Heaven of Pre-Ch'in Thought," in *The New Asia Journal*, II-2. Hong Kong, 195.
_____ . "The Development of Ideas of Heavenly Mandate Since the Ch'in Dynasty," in *The New Asia Journal*, VI-2. Hong Kong, 1964.
T'ang Hsien-tsu. "Han-tan chi," in *T'ang Hsien-Tsu chi*. Shanghai: Chung-hua shu-chü, 1962.
_____ . *Mu-tan t'ing*. Taipei: Wen-kuang ch'u-pan-she, 1974.
T'ao Shu. *T'ao Ching-chieh chi chu*. Hong Kong: T'ai-p'ing-yang ch'u-pan-she, 1964.
Taylor, John Russell. *Anger and After: A Guide to the New British Drama*. London: Methuen & Co., 1963.
Thompson, David. *Dante's Epic Journey*. Baltimore & London: The John Hopkins University Press, 1974.
Thompson, Laurence G. *Chinese Religion: An Introduction*. Belmont, California: Dickenson Publishing Co., 1969.
Tien Shu-chuang. Ed. *History of Chinese Modern Literature*. Ch'angch'un: chi-lin Jen-min ch'u-pan-she, 1957.
Tillyard, Eustace Mandeville Wetenhall. *Mythical Elements in English Literature*. London: Chatto & Windus, 1961.
_____ . *Myth and the English Mind: From Piers Plowman to Edward Gibbon*. New York: Collier Books, 1962.
Tindall, W.Y. *James Joyce: His Way of Interpreting the Modern World*. New York: Charles Scribner's Sons, 1950.
Ting Ying. *Chung-kuo shang-ku shen-hua ku-shih*. Hong Kong: Shang-hai shu-chü, 1960.
Ts'ao Hsüeh-ch'in. *Hung-lou meng*. Tainan: Wang-chia ch'u-pan-she, 1975.
Tu John Er-wei. "A Comparative Study of the Astrological Myths of China and Those of Babylonia," in *Bulletin of the College of Arts*, 1966. Taipei: National Taiwan University Press, 1966.
_____ . "The Meaning of K'un Lun Myths," in *The Contemporary Thought Quarterly*, I-1. Taipei: 1961.
_____ . *The Mythological System of the Mountain Sea Classic*. Taipei: Hua-ming shu-chü, 1960.
_____ . *The Religious System of Ancient China*. Taipei: Hua-ming shu-chü, 1951.
_____ . *Studies About the Religions of Ancient China*. Taipei: Hua-ming shu-chü, 1959.
Tu Shih-chieh. *Hung-lou meng pei-chin tao-yü shih-k'ao*. Taichung: Author, 1971.
Unterecker, John. *A Reader's Guide to William Butler Yeats*. New York: Noonday Press, 1959.
Vickery, John B. *Myth and Literature: Contemporary Theory and Practice*. Lincoln: University of Nebraska Press, 1966.
_____ . *The Literary Impact of the Golden Bough*. Princeton: Princeton University Press, 1973.
Waley, Arthur. Tr. *Monkey*. Penguin Books, 1973.
Wang Chi-chen. Tr. *Dream of the Red Chamber*. New York: Twayne Publishers, 1958.
Wang Hsiao-lien. "Studies about Titan K'ua Fu," in *The Continent Magazine*, 46-2. Taipei, 1973.

Wang Kuo-wei. "Criticism on the Dream of the Red Chamber," in *Ku-tien wen-hsüeh yen-chiu tzu-liao hui-pian*, I-3. Ed. Yi Li, Peking: Chung-hua shu-chü, 1963.
Wang Meng-ou. "Astrology and Sorcery of the Yin-yang School," in *Bulletin of the Institute of History and Philology, Academia Sinica*, XLIII-3. Taipei:, 1971.
_____. *Modern Annotation and Translation of the Book of Rites*. Taipei: Shang-wu yin-shu-kuan, 1971.
Wang Shu-ching. *The Biography of Lu Hsün*. Shanghai: Lien-he ch'u-pan-she, 1947.
Watson, Burton. Tr. *The Complete Works of Chuang Tzu*. New York: Columbia University Press, 1068.
Watt, Ian. *The Rise of The Novel*. Penguin Books: 1972.
Watts, Alan Wilson. *Cycle in Buddhist Philosophy*. Phonotape, Toronto, CBC, 1970.
_____. *Myth and Ritual in Christianity*. London: Thames & Hudson, 1954.
Wayman, Alex. "No Time, Great Time, and Profane Time in Buddhism," in *Myth and Symbols: Studies in Honour of Mircea Eliade*. Ed. Joseph M. Kitagawa and Charles H. Long. Chicago & London: The University of Chicago Press, 1969.
Wayne, Philip. Tr. *Faust: Part II*. Middlesex: Penguin Books, 1959.
Wei Cheng-t'ung. *A Critical Approach to Chinese Culture*. Taipei: Shui-niu ch'u-pan-she, 1969.
_____. *A Critical Approach to Chinese Philosophy and Thought*. Taipei: Shui-niu ch'u-pan-she, 1968.
Weinstein, Norman. *Gertrude Stein and the Literature of the Modern Consciousness*. New York: Frederick Ungar Pub. Co., 1970.
Wei T'ien-t'ung. "Chung-kuo ku-tai te shen-hua ching-shen," in *Chung-wai wen-hsüeh*, III-8. Taipei, 1975.
Weisinger, Herbert. *The Agony and the Triumph: Paper on the Uses and Abuse of Myth*. East Lansing: Michigon State University Press, 1964.
Wellek, René and Austin Warren. *Theory of Literature*. New York: Harcourt, Brace and Co., 1942.
Welzels, Water D. Ed. *Myth and Reason: A Symposium*. Austin & London: University of Texas Press, 1973.
Wen I-tuo. *Shen-hua yü shih*. Peking: Chung-hua shu-chü, 1956.
West, R. Hunter. *Milton and the Angels*. Athens: University of Georgia Press, 1955.
Wheelwright, Philip Ellis. *Metaphor and Reality*. Bloomington: Indiana University Press, 1962.
_____. *The Hidden Harmony: Essays in Honour of Philip Wheelwright*. New York: Odyssey Press, 1966.
Wilhelm, Hellmut. *Change: Eight Lectures on the I Ching*. Tr. Cary F. Baynes. New York: Harper & Row Publishers, 1960.
_____. *Heaven, Earth, and Man in the Book of Changes*. Seattle & London: University of Washington Press, 1977.
Wilhelm, Richard. *The I Ching: Book of Changes*. Tr. Cary F. Baynes. New Jersey: Princeton University Press, 1967.
Willey, Basil. *The Seventeenth Century Background*. New York: Doubleday & Co 1953.
Wittgenstein, Ludwig. *The Blue and Brown Books*. Oxford: Basil Blackwell, 1958.
Woodcock, George. "Songs of the Siren: Notes on Leonard Cohen," in *Odysseus Ever Returning*. Toronto: McClelland and Stewart, 1970,
Wu Ch'eng-en. *Hsi-yu chi*. Tainan: Wang-chia ch'u-pan-she, 1975.
Wu Hung-yi. "The Tragic Spirit in the Dream of the Red Chamber," in *Essays on the Dream of the Red Chamber*. Taipei: Yu-shih ch'u-pan-she, 1972.
Yu Kuo-en. *Ch'u-tz'u lun-wen chi*. Shanghai: Wen-yi lien-he ch'u-pan-she, 1959.
Yü Ying-shih. "Life and Immortality in the Mind of Han China," in *Harvard Journal of Asiatic Studies*, 25. 1964-65.
Yüan Ke. *Chung-kuo shang-ku shen-hua*. Peking: Chung-hua shu-chü, 1960.

INDEX

A

A Dream in Han Tan Inn 5, 92, 98, 101
A Vision 6, 148
Abraham 127
Amour 85
archetypal image 121
  image of awakening or resurrection 123
  image of conversion or canonization 123
  image of spring or autumn 122
  image of the garden 122
  image of the heroine 123
  image of the mountain 121
  image of the recluse 123
  image of water 122
  images of dream and death 123
archetype of transfiguration and reconciliation 99
Astarte 128
Augustine 130
Axis Mundi 129

B

Baal 128
Bacon, Francis 138, 153
Bamboo Books, The 33
Beatrice 131, 135
Beautiful Losers 7, 169, 171, 173, 175, 176
Beckett, Samuel 7, 163, 166, 180, 187
Biography of Ah Q, The 7, 180, 185
Book of Changes, The 50, 51, 55-57
Book of Documents, The 33
Book of Odes, The 78
Book of Rites, The 37
Byron 147

C

Cartesian detachment 154
Catholicism 130
Celestial Journey of the Great Man 85
Cerberus 132
Chekhov, Anton 7, 180, 183
Chinese Buddhism 69
  approach to the phenomenal 70
  Bodhidharma 71
  Chih I 70
  Ch'an 70
  Ch'an movement 71
  concept of dharma 72
  Fa Tsang 70
  Heart Sutra 107
  Hua-yen School 70
  Hui Yüan 69
  Hung Jen 71
  Identification of Man with the Universal Mind 69
  Indian Buddhism 72
  Karma 69
  Mahayana Buddhism 107
  Nirvana 69
  Pure Land School 69
  Samsara 69
  Tao Sheng 69
  Ten Characters of Thusness 70
  Three Levels of Truth 70
  T'ien-t'ai School 70
  Wheel of Life and Death 69, 72
  Zen School 70
Chinese cosmology 73
Chinese ritual 40
  Altar to Heaven 42
  Ancestor Worship 48
  anthropocentrism 48
  Book of Changes, The 50, 195
  cannibalistic and sexual assaults 109
  Chiao 41, 110
  concept of "Te" 49
  cyclic basis of the rite 43
  Duke of Chou 55
  Earth Sacrifice 42
  Eight Trigrams 50, 53
  Feng Shan Ritual 84, 110
  Feng Shan Sacrifice 44
  Hall of Light, the 44, 45
  impersonalization of Heaven 49
  Institution of the Hall of Light 44
  King Wen 55
  Lower Palaeolithic and Neolithic Ages 40
  mana 109
  Mandate of an Impersonal Heaven 49
  Ming t'ang 45, 84, 110, 193
  multiplication of the eight trigrams 55
  Omnipotence of T'ien 49
  oracle bones 48

210 Index

Pa Kua 50, 53
peasant omen 50
Pill of Divinity 110
Primal Arrangement of the Pa Kua 52
psychological and ethical basis of the Ancestor Worship 49
Sacrifice to Earth 44
Sacrifice to Heaven 42, 44
Sacrifice to Heaven and Earth 41
Sacrifice to the Cosmic Mountain 44
Sacrifice to the Four Directions 43
Sequence of Later Heaven 55
six jade offerings 43
sixty-four hexagrams 55
structure of Ming T'ang 47
Suburban Sacrifice 41
summer solstice 42
Supreme Being 48
Ti 48
totemism 40
winter solstice 42
Chiu Ko 83
Christian Advent 129
Christian Aristotelianism 132
Christian faith 127
Christian liturgical year 129
Christmas 129
Ch'ü Yüan 5, 78
Ch'uan Ch'i 91, 98, 102
Clarissa 158
Classical Walpurgisnacht 143
Cogito ergo sum 156
Cohen, Leonard 7, 169
Coleridge 145
Commedia, The 135, 197
Concept of change 55
Condillac 155
Confessions 141
Confucianism 62, 109, 195
  Chou Tun-i 63
  Chu Hsi 63
  Chung Yüng 62
  Ch'eng I 63
  Confucius 62
  Empirical School of Confucianism 63, 118
  Golden Mean 62
  humanism 62
  Idealistic School of Mind 63
  Lo Ju-fang 92
  Lu Hsiang-shan 64

Neo-Confucianism 63, 107
operation of Yin and Yang 63
Rationalistic School of Principle 63
Tai Chen 64
Three Reigns 62
Tung Chung-shu 62, 65
unity of man with cosmos 62
Wang Yang-ming 64, 92, 107
Yin-yang Confucianism 62
cosmic mountain 121
  Holy Mountain of Buddha 103
  Mount Chao 14, 111
  Mount Ch'ang Liu 22
  Mount Cool Wind 15, 81
  Mount Five Elements 105
  Mount Flower and Fruit 103
  Mount K'un Lun 14, 22, 24, 36, 44, 80, 84, 88
  Mount of Fire 15, 24
  Mount Teng Pao 14
  Mount T'ai 16, 44
  passageways to Heaven 14
  Pillar of the Universe 22
  spiral roads 14
  Sun-moon Mountain 19
cosmic transformation 122
cosmic tree
  Chien Tree 15, 16
  Divine Mulberry Tree 21
  Fruit of Life 109
  Fu Sang Tree 20, 21, 81
  Hanging Garden 15
  home of magical trees 15
  Peach Garden 109
  Peach Tree of Immortality 35, 104
  Sacred Laurel Tree 25
  Sacred Peach Tree 20
  Tree of Knowledge 129
  Tree of Life 122, 132
  Tree of the Cross 129
cosmic water 122
  Abyss of Sinking Water 15, 24
  Abyss Yü 20, 35
  Crescent of Sinking Sand 22, 82
  Dark-River 22
  Great Marshes of Chaos and Life 35
  Lake of Purity 15, 25, 81
  River of Clearness 15
  Spring of Sorrow 20
  Valley Meng 25
  Valley of Chaos 20
  Valley T'ang 20

Index

Country of Splendid Flowers 14
Cross of Christ 129
Cyclic myth 3, 72
  absolute totality 73
  as index of culture 193
  celestial progress to divinity 103
  Chinese cyclic myth 4
  cyclic aspects of Confucianism 62
  cyclic mentality 193
  cyclic schema 121, 197
  development of the cyclic concept 73
  emergence of the God of Time 34
  evolution of the cyclic myth 77
  fall into linearity 138
  harmony with nature 68
  immortality 35
  in Chinese mythology 9
  incessant pursuit from limbo to resurrection 92
  lunar symbol 73
  mana 73
  myth in natural symbolism 130
  Myth of Divine Administration 32, 38
  Myth of Hsi He 34
  Myth of the Eternal Becoming 37
  Myth of the Reconciliation with Time 36
  Myth of the Rise of Time and Mortality 32
  Myth of the Ten Suns 34
  Myth of Titan P'eng Tsu 34
  mythic quest for the unity with cosmos 79
  Myths of the Rebellion Against Time 35
  ordeal descent into eternity 98
  pastoral quest for harmony with nature 86
  periodic regeneration of man and cosmos 65
  poet's vision at the cycle's centre 144
  portrayals of the associative and the sensational 159
  prophecy about cyclic cosmos 148
  quest cycle 196
  quest of the hero 5
  reconciliation of man and nature 90
  reconciliation of man and time 68
  sensory details of successive moments 156
  sentimental moments with intense feeling 158
  spiral transcendence to the centre of circles 131
  spiritual quest for eternal return 87
  temporal envisagement 32
  terrestrial journey to disillusionment 111
  the way of nature 73
  T'ien Jen Ho I 4
  unity of man and cosmos 4, 100, 126
  Viconian myth of the course of humanity 149
cyclic myth of galatea 143
cyclic ontology 4, 193, 194
cyclic symbolism 127

D

Dante, Alighieri 6, 131
Darwin, Charles 181
Defoe, Daniel 6. 156
Descartes, René 138, 194
Divine Commedia 6, 131
Dramatic Romance 91, 98
Dream of the Red Chamber, The 5, 102, 111, 117

E

Earth Spirit 143
Easter 129
Easter Sunday 135
Easter Thursday 135
Eden 129, 136
Eisenstein 169
Eliade, Mircea 106, 128
Elizabethan literature 138
Empyrean 135
Encountering Sorrow 5, 79
Epiphany 129
Epistle 128
Eros 143
Ezekiel 127

F

Faerie Queene, The 138
Fallen Adam 139
Faust 6, 110, 141, 142
Fichte, Johann G. 145
Fielding, Henry 156, 159, 179

Fielding, Henry 156, 159, 179
Finnegans Wake 6, 149-151, 177
first millennium 131
Four Dream Plays 91
Frye, Northrop 78, 111

G

Garden of Great Wonder 6, 118
geistesgeschichte 7, 179
Geryon 132
goddesses
  Buddhisattva of Mercy 105
  Ch'ang Hsi 20
  Goddess Ch'ang O 23, 24, 26, 36, 44
  Goddess of Dark Mystery 18
  Goddess of Disillusionment 112
  Goddess of Matchmaking 80
  Goddess of Music and Marriage 12
  Goddess of the Moon 20
  Goddess of the Sun 20
  Hsi He 20
  Hsi Wang Mu 24, 25, 81
  Mother of Mothers 12
  Nü Hsü 80
  Nü Wa 11-13, 111
  Proserpina 132
  Supernal Mother Goddess 13, 24, 81, 111
gods
  Archer God Hou I 24, 110
  Buddha 104
  Chaos 9, 10
  Chief of the Wind 17
  Chien Ti 81
  Chu Jung 13, 28, 31
  Chü Lung 26, 31
  Chü Mang 27, 31
  Chuan Hsü 18, 19, 29, 30, 38, 79
  Ch'ung 18, 27, 31, 81
  Daybreak 15
  Diamond Angels 106
  Dragon King 104
  Eight Taoist Immortals 98
  Elder of the Rain 17
  Fu Hsi 12, 45, 51
  God of Earth 19, 22, 27
  God of Fire 13, 28, 111
  God of Gold or Metal 29
  God of Growth 28
  God of Hades 27, 35
  God of Harvest 29
  God of Hibernation 29
  God of Life 18, 27
  God of Punishment 29
  God of Soil 27
  God of Spring and Life 81
  God of the Autumn 29, 43
  God of the East 19
  God of the Great Ground 28
  God of the Multitude 28
  God of the North Sea 38
  God of the Spring 27, 43
  God of the Summer 28, 43
  God of the Winter 29, 43
  God of Time 19, 20
  God of War 15
  God of Water 13, 29, 38, 111
  God of Wood 27
  Hercules 132
  hierophany 39
  Hou I 24
  Hou T'u 22, 26, 31
  Hsüan Meng 29, 38
  Huang Ti 14, 26, 30
  I Ming 19, 34
  Jade Emperor 106
  Ju Shou 28, 31
  Judge of Metempsychosis 93
  Kai 28, 31
  K'ai Ming 15
  Kung Kung 13
  Li 18, 28, 31
  Li Chu 16
  Ling Fen 82
  Lord of Yü 81
  Lu Wu 16, 81
  P'an Ku 9, 33
  Prince Sao K'ang 81
  Shao Hao 28, 30
  Shen Nung 45
  Sovereign God of the East 27, 43, 81
  Sovereign God of the North 29, 38, 43
  Sovereign God of the South 28, 43
  Sovereign God of the West 28, 43
  Supremacy of Heaven 104
  Supremacy of the Universe 14, 26
  Supreme Being 26, 79
  T'ai Hao 27, 30
  Ten Judges of Death 104, 106
  Theseus 132
  Ti Chün 19
  Titan Ch'ih Yu 14, 17, 110

Titan K'ua Fu 18, 106
Titan P'eng Tsu 82
Vajrapanis 106
Wu Hsien 81
Wu Kang 25
Yen Ti 17, 28, 30
Yin Chao 15
Yü Ch'iang 29, 31, 38
Yüan Meng 29, 31, 38
Goethe, Johann Wolfgang von 6, 141
Gold-vase Plum, The 102
Golden Age 13, 33, 88
Golden Bough, The 151
Golden Pheasant 20
Good Friday 135
Gorgon 132
Great Wheel, the 148, 177
Gretchen 143
Griffin 136

H

Hades 96
Hall of Light, the 6

Hanging Gardens 25
Han-tan meng chi 5, 92, 98
Happy Days 7, 180, 187
Hebraic Messianism 127, 193
Hebrews 128
Hegel, Georg 179
Helen of Troy 144
Historicity of Christ 131
Holy Trinity 135, 137
Homecoming, The 7, 180, 190
homogeneity 198
House of the Spring 81
Hsi-yu chi 5, 103
Hua Hsü Country 14
Hume, David 154
Hun Tun 10
Hung-lou meng 5, 111
Hunt, Theodore 179

I

I Ching 50
Ibsen, Henrik 7, 180, 181
Inferno 135

J

Jade Pillow 99
Jeremiah 127
Jesus 128
Job 127
Joseph Andrews 156
Journey to the Mothers 144
Journey to the West, The 5, 102, 103, 106, 109
Joyce, James 6, 148, 149
Judaism 127
Judaeo-christianity 193
Jung, Carl 96, 99

K

Kant, Immanuel 145, 181
Kuei Ch'ü Lai Tz'u 5, 86

L

Last Judgement 140
Lent 129
Li sao 5
Limbo 93, 96
linear eschatology 127, 193, 194
Liu Yi-ch'ing 99
Lo Ju-fang 5, 95, 97
Locke, John 154, 160
Logos 50
London theosophists 148
Lord Yahweh 127
Lu Chi 86
Lu Hsün 7, 180, 185
Lucifer 132, 135

M

macrocosm 50, 195
Mandate of Heaven 49
Marx, Karl 181
Mephistopheles 142, 144
Michael 140
microcosm 50, 195
Middle Ages 194
Milton, John 6, 138, 139
Ming t'ang 6, 44
Minos 132
Minotaur 132
Mithraism 135

Moll Flanders 6, , 156, 157, 162
Montage 169
Montaigne, Michel de 138
Moses 110
Mountain and Sea Classic, The 33, 37
Mu-tan t'ing 5, 92
Mutability Cantoes 138
myth 1-3
  displacements 2
  Eliade, Mircea 1, 4
  Feder, Lillian 2
  Frye, Northrop 2, 3
  Journey of the Sun 19
  Jung, Carl 1
  Langer, Susanne 2
  Malinowski, Bronislaw 1
  Myth of Archer God Hou I 23
  Myth of Disasters on Earth After the Estrangement 19
  Myth of Divine Administration 193
  myth of Eden 129
  myth of Mount K'un Lun 129
  Myth of Paradise Rebuilt 13
  Myth of Passageways to Heaven 14
  Myth of the Creation 9
  Myth of the Creation of Man 11
  Myth of the Divine Administration 26
  Myth of the Divine Conference 16
  Myth of the Estrangement of Earth from Heaven 18
  Myth of the Flight to the Moon 24
  Myth of the God of Time 19
  Myth of the Great Deluge 13
  Myth of the Rebellion of Titan Ch'ih Yu 16
  Myth of the Silver Age 14
  Myth of Titan K'ua Fu 22
  Myth of Titan P'eng Tsu 21
  reconstruction of fifteen Chinese Myths 9

N

Neo-Confucian thought 92
Neo-Platonism 148
New Creation 130
Nietzsche, Friedrich 179, 181, 183
Nine Songs, The 83
Nine Underworlds 82
Ninefold Heaven 18, 80, 82
Northern Play 91

O

Ode To Dejection 145
Ode to the West Wind 6, 144, 145
Oedipus 103, 105, 110
omnigoodness 49
omnijustice 49
omnipotence 49
oracle bones 38
Oriental 197
Origin of the Species 181

P

Palace of the Moon 25
Pamela 6, 158, 162,
Paradise Lost 6, 138-140, 197
Passiontide 129
Pentecost 129
Peony Pavilion, The 5, 92, 94, 95, 101, 110
Perseus 132
Peter 128
Pietas 85
Pillar of the Universe 13
Pinter, Harold 7, 180, 190
Pluto 132
Pope, Alexander 179
Prelapsarian Adam 139
Primal Arrangement 51
Primum Mobile 132
Prometheus 110
Ptolemaic order 132
Purgatory 132

R

Rabelaisian mockery 5, 110
racial memory 144
Rebellion of Ch'ih Yu 33
Remarks Concerning the States, The 33
Renaissance 6, 138, 194
Return, The 86, 87, 90
Rhapsody of Literature 86
Rhapsody of the Goddess, The 85
Rhapsody of the Nymph of River Lo, The 85
Rhapsody on the Return 5
Richardson, Samuel 6, 158
ritual 40
Robinson Crusoe 159

Romance of the Three Kingdoms, The 102
Romantic movement 144
Rousseau, Jean J. 141, 155

S

Sacred Jade Cock 20
Satan 139
Scholars, The 102
Seagull, The 7, 180, 183
Second Coming 149
sensus spiritualise 83
Shakespeare, William 138
Shelley, Percy Bysshe 6, 144, 145
Shen Chi-chi 99
Sleeping beauty 97
Songs of the South, The 78
Son's Anointment 139
Sorge 144
Southern Show 91
Spengler, Oswald 148
Spenser, Edmund 138
Spring and Autumn Annals of Lü, The 37
Ssu-ma hsiang-ju 85
St. Augustine 128
Sterne, Laurence 6, 159
story cycles 102
striving for meaningful duration 141
succession 152
Sung Yü 85
superman 181
symbolic phases 124
　call to the quest 124
　cosmic progress 85
　itineraria 79, 80, 131
　love quest 85
　progress 84
　progress of the cosmic journey 83
　quest 79, 83, 124
　quest of the goddess 85
　reconciliation 125
　return of the hero 125, 131
　tristesse 79
　tristia 79, 83, 86, 87, 131
symbolism of beauty and the feminine 84
symbolism of vegetable adornment 83
synchronicity 96

T

Tales of the Fantastic 91, 102
T'ang Hsien-tsu 5, 90, 91, 95, 98, 110, 111
Tao 49
T'ao Ch'ien 5, 86
Taoism 66, 195
　Book of Chuang Tzu, The 67
　Book of Huai-nan Tzu, The 67
　Chuang Tzu 66, 68
　cyclic process as the Way 67
　harmony of man with nature 66
　Lao Tzu 66, 68
　Liu An 67
　Neo-Taoists 68
　non-action 66
　non-being 68
　philosopher of changes 66
　Tao 66
　Te 66
　wu 68
　wu-wei 66
Ten Suns 23
theory of spontaneous association 160
T'ien Jen Ho I 4, 126
time 1, 4
　Chinese conception of time 77
　Chinese historiography 75, 76
　Chinese temporal thought 74, 76
　co-existence of past, present and future 153
　continuity history writing 75
　cyclical cosmic time 74
　duration 1, 152, 153, 156
　eighteenth century 154, 156
　French temporal thought 155
　Golden Age 75
　Great Peace 76
　great time 106
　Great Togetherness and Harmony 76
　linear historical time 75
　memory 153
　no time 106
　profane time 106
　realizable utopia 75
　Renaissance 153
　rise of the English novel 6
　seventeenth-century 154
　succession 1, 152, 156
　Ta T'ung 76

T'ai P'ing 76
the Infinite 106
time 152
timeless 152, 153
World Age 75
time spirit 7
Tom Jones 179
Tristram Shandy 6, 159
Troilus and Cressida 138
Ts'ao Chih 85
Ts'ao Hsüeh-ch'in 5
Twenty Poems After Drinking Wine 88, 89

U

Ulysses 138, 149, 150
Underworld of Darkness 106
Unger, Rudolf 180
Unnameable, The 164

V

Variety Play 90
Vico 148
Vico Road 193
Viconian philosophy 150
Vico's theory 149, 150
Virgil 135

W

Wang Kuo-wei 118
Water Margin, The 102
Watt 7, 166, 176
Western Paradise 106, 112
Wheels and Butterflies 148
Wild Duck, The 7, 180, 181
Winter's Tale, The 138
Wittgenstein, Ludwig 7, 166, 188
World Inside the Pillow, The 99
World-soul 142, 144
World-view 196
Wu Ch'eng-en 5, 103
Wu Chi Door to Heaven 19

Y

Yahweh's wrath 128
Yeats, William Butler 6, 148
Yin and Yang 11

Yin-yang School 60, 195
 Book of Documents, The 60
 Book of Rites, The 60
 doctrine of the Five Elements 60
 dual-multiple cosmogony 60
 Grand Norm 60
 Monthly Observances 60
 penta-cyclical cosmology 60
 Spring and Autumn Annals of Lü, The 61
 Tsou Yen 60, 61
 Yin-yang concept 60
 Yin-yang theory 55, 61
Yü-chou 74

ASIAN THOUGHT AND CULTURE is designed to cover three inter-related projects: (1) *Asian Classics Translation* (including those modern Asian works that have been generally accepted as "classics"), with notes and commentaries provided; (2) *Asian and Comparative Philosophy and Religion*, including excellent and publishable Ph.D. dissertations, scholarly monographs, or collected essays; and (3) *Asian Thought and Culture in a Broader Perspective*, covering exciting and publishable works in Asian culture, history, political and social thought, education, literature, music, fine arts, performing arts, martial arts, medicine, etc.

The series editor is:
Charles Wei-hsun Fu
Department of Religion
Temple University
Philadelphia, PA 19122